PROGRESS ON PEST MANAGEMENT IN FIELD VEGETABLES

The Experts' Group Meeting on 'Integrated Plant Protection in Field Vegetables' was organized jointly by:

 Commission of the European Communities
Directorate-General Agriculture
Coordination of Agricultural Research

 International Organization for Biological and Integrated Control
West Palaearctic Regional Section
Working Group Integrated Control in Brassica Crops

*Proceedings of the CEC / IOBC Experts' Group Meeting / Rennes
20-22 November 1985*

PROGRESS ON PEST MANAGEMENT IN FIELD VEGETABLES

Edited by
R.CAVALLORO
Commission of the European Communities

C.PELERENTS
International Organization for Biological and Integrated Control

Published for the Commission of the European Communities by
A.A.BALKEMA / ROTTERDAM / BROOKFIELD / 1988

The texts of the various papers in this volume were set individually
by typists under the supervision of each of the authors concerned.

Publication arrangements: *P.P.Rotondó*, Commission of the European Communities,
Directorate-General Telecommunications, Information Industries and Innovation, Luxembourg

LEGAL NOTICE
Neither the Commission of the European Communities nor any person acting on behalf of
the Commission is responsible for the use which might be made of the following information.

EUR 10425

Published by
A.A.Balkema, P.O.Box 1675, 3000 BR Rotterdam, Netherlands
A.A.Balkema Publishers, Old Post Road, Brookfield, VT 05036, USA

ISBN 90 6191 759 X

Foreword

This publication includes both the texts of papers given and the opening addresses and final conclusions of the experts' meeting held at Rennes, France, from 20 to 22 November, 1985, on "Integrated Plant Protection in Field Vegetables". It was organised jointly by the Commission of the European Communities and by the International Organisation of Biological and Integrated Control.

It contains forty contributions, divided into specific sessions, which deal with very important crops such as cabbages, carrots, tomatoes, potatoes, beans and artichokes.

There is no doubt that vegetable crops, many of which are eaten raw, deserve particular attention concerning the maximum reduction of the use of pesticides, with the adoption of means and control strategies against pests and diseases.

These strategies are based on the prediction of attacks, the knowledge of tolerance thresholds, the selection of resistant cultivars, the growing of intercalated crops and the use of parasites, predators, microorganisms or other biological, cultural or bio-technical methods.

Close collaboration between researchers allows us to attain these aims more quickly, as has been shown above all for the prediction of attacks of pests based on the analysis of samplings performed for several seasons in different countries using the same protocol.

It is also encouraging to note that integrated control methods have gone beyond the laboratory and experimental field stage. It is applied by vegetable cultivators, as especially in the growing of cauliflowers and Brussel sprouts.

For a better knowledge of the ecology, behaviour and ethology of pests new protection techniques are being explored, such as the use of traps and sexual pheromones, and recourse to cultural methods.

Thirthy-four experts, from twelve countries and two international organisations, took part in this meeting. They presented and discussed the results of their research on vegetable crops, for efficient protection against pests and diseases. They feel it is possible to put into practice a correct application of integrated pest control, with safe means, methods and strategies of intervention.

R. Cavalloro, C. Pelerents

STRUCTURE OF THE MEETING

Scientific Committee

Brunel E., I.N.R.A. - Centre de Recherches de Rennes, F-Le Rheu
Cavalloro R., C.E.C. - Joint Research Centre, I-Ispra
Coaker T., Department of Applied Biology - University, GB-Cambridge
Pelerents C., Faculteit van de Landbouwwetenschappen - Rijksuniversiteit, B-Gent

Sessions' Organization

Introduction

Opening address by R. Cavalloro and C. Pelerents

Session 1

Brassica crops   Chairman : T. Coaker

Session 2

Carrot crops   Chairman : R. Cavalloro

Session 3

Tomato crops   Chairman : E. Brunel

Session 4

Other crops and general aspects
on pest control in field vegetables   Chairman : C. Pelerents

Conclusions

General discussion, closing remarks, and recommendations by
Scientific Committee

Local Secretariat

Brunel E.
Institut National de la Recherche Agronomique - Centre de Recherches de Rennes - Domaine de la Motte-au-Vicomte, Le Rheu

Proceeding Desk

Rotondó P.P.
CEC - Directorate-General Telecommunications, Information, Industries & Innovation, Luxembourg

Table of contents

Introduction

Opening address by the Commission of the European Communities representative 3
R.Cavalloro

Opening address by the International Organization for Biological and Integrated 7
Control Representative
C.Pelerents

Session 1: *Brassica crops*

Problems in monitoring cabbage root fly, *Delia radicum (brassicae)* by oviposition 11
traps
H.den Ouden

Use of egg-traps in a warning system against the cabbage root fly (*Delia radicum*) 15
B.Bromand

Thermal requirements for cabbage root fly, *Delia radicum*, development 21
R.H.Collier & S.Finch

Factors regulating diapause and aestivation in the cabbage root fly, *Delia radicum* 27
R.H.Collier & S.Finch

Emergence of cabbage root flies from puparia collected throughout northern Europe 33
S.Finch, B.Bromand, E.Brunel, M.Bues, R.H.Collier, R.Dunne, G.Foster, J.Freuler,
M.Hommes, M.Van Keymeulen, D.J.Mowat, C.Pelerents, G.Skinner, E.Städler &
J.Theunissen

Preliminary studies of the factors regulating late emergence in certain cabbage root 37
fly, *Delia radicum*, populations
R.H.Collier & S.Finch

Mortality of the immature stages of the cabbage root fly 45
S.Finch & G.Skinner

Forecasting the times of attack of *Delia radicum* on brassica crops: Major factors 49
influencing the accuracy of the forecast and the effectiveness of control
S.Finch & R.H.Collier

Experiences with action thresholds in cabbage crops in the Federal Republic of 55
Germany and first results of a collaborative field trial
M.Hommes

Oilseed rape crops as a source of cabbage root fly infestations for cruciferous 61
vegetable crops
G.Skinner & S.Finch

Analysis of cabbage root fly population measurements and the derivation of control 67
decisions
M.Van Keymeulen & C.Pelerents

Effect of chemical treatment on *Delia radicum* and its parasitoids *Aleochara* 75
bilineata and *Trybliographa rapae*
B.Bromand

Development of various formulations for controlled release of naphthalene as a 83
repellent against oviposition of the cabbage root fly, *Delia radicum (brassicae)*
H.den Ouden

The control of cabbage root fly, *Delia brassicae* (Wied.), in transplanted brassicas by 87
insecticide application to the seedbed – A report of collaborative work
D.J.Mowat

The application of insecticides for cabbage root fly (*Delia radicum*) control in 95
cabbage grown in small peat modules
R.Dunne & J.Coffey

Investigations of the resistance of cabbage cultivars and breeders lines to insect pests 99
at Wellesbourne
P.R.Ellis & J.A.Hardman

Sequential sampling of insect pests in Brussels sprouts 107
J.Theunissen

Conditions for implementation of supervised control in commercial cabbage 117
growing
J.Theunissen & H.den Ouden

Preliminary observations on swedes with resistance and susceptibility to turnip root 123
fly (*Delia floralis*) in Scotland
N.Birch

Problems encountered with the use of pheromone traps for monitoring *Mamestra* 129
brassicae populations in Belgium
C.Pelerents & M.Van de Veire

Monitoring of *Plutella xylostella* adults by pheromone traps – A report of a 135
collaborative trial
J.Theunissen, J.Freuler, M.Hommes, D.J.Mowat, F.Van der Steene & F.Gfeller

Detection of movements of oligophagous Lepidoptera between host plant sowings   137
by means of sexual traps with specific reference to *Plutella xylostella* and *Brassica*
species
R.Rahn

Studies of soil receptiveness to clubroot caused by *Plasmodiophora brassicae*:   145
Experiments on responses of a series of vegetable soils in Brittany
F.Rouxel, M.Briard & B.Lejeune

Session 2: *Carrot crops*

Field and laboratory studies on the behaviour of the carrot fly, *Psila rosae*   155
G.Skinner & S.Finch

Monitoring of the carrot rust fly, *Psila rosae*, for supervised control   161
H.den Ouden & J.Theunissen

Towards an effective and economic carrot fly trap: Progress report 1981-1985   167
E.Städler, F.Gfeller, K.Keller & H.Philipsen

Integrated pest management in Danish carrot fields: Monitoring carrot fly (*Psila*   169
rosae F.) by means of yellow sticky traps
H.Philipsen

Integrated pest management in Danish carrot fields: Monitoring of the turnip moth   177
(*Agrotis segetum* Schiff., Lepidoptera: Noctuidae)
P.Esbjerg

Control of *Heterodera carotae* Jones (1950): Effectiveness of available methods and   187
prospects
M.Bossis

Session 3: *Tomato crops*

Pest problems in field tomato crops in Spain   197
R.Albajes, R.Gabarra, C.Castañe, E.Bordas, O.Alomar & A.Carnero

Development of a control programme against the aphids (*Macrosiphum euphorbiae*   209
Thomas and *Myzus persicae* Sulzer) in processing tomato cultures in the south-east
of France
R.Bues, J.F.Toubon & H.S.Poitout

The present status on bacterial diseases of tomatoes in Greece   221
C.G.Panagopoulos

Disease resistance: An excellent resource for world-wide biological control in   223
tomato
M.Cirulli & F.Ciccarese

Session 4A: *Other crops*

Population dynamics of aphids as vectors of potato viruses in Portugal   257
M.O.Cruz de Boelpaepe & M.I.Rodrigues

Susceptibility of various bean cultivars to bacterial blights   275
C.G.Panagopoulos & D.A.Biris

Integrated control of the artichoke moth (*Gortyna xanthenes* Ger.) (Lepidoptera: 277
Noctuidae) in Italy
E.Tremblay & G.Rotundo

Session 4B: *General aspects on pest control in field vegetables*

Insect pest management by intracrop diversity: Potential and limitations   281
T.H.Coaker

Present status of supervised pest control methods in field vegetables   289
J.Freuler

Conclusions and recommendations   295

List of participants   297

Index of authors   302

Introduction

Opening address by the Commission of the European Communities representative

R.Cavalloro
Principal Scientific Officer CEC

Ladies and Gentlemen,

Field vegetables occupy a preeminent place among crops, which are now being available throughout the year. Market demand is always high and of extremely relevant economic value.

From Eurostat data of 1984, in the member countries of the European Communities, the production of only fresh vegetables reached about 30 million tons, without considering potatoes which alone reached the same production figure. Soon, with the foreseen enlargement to twelve countries with the adhesion of Spain and Portugal, 40 million tons will be produced, of which one-fourth is due only to tomatoes.

The request for healthy products and of high quality, demands even more from producers the use of all the agricultural practices most suitable, and the cultural techniques which fulfil this aim.

Among these the protection of crops against pests and diseases occupies a place of fundamental importance. This occur throughout the phenological development of the plant, from sowing until harvest.

The list of animal pests and of diseases which may attack field vegetables, often with disastrous results for the crop itself, is impressive, and special care should be taken to act, if such is the case, quickly, and with the correct choice of control means, products, and strategies which certainly defend the crop, but which, at the same time, do not cause undesirable secondary effects.

The awareness of a rationalisation of the phytosanitary measures has lead to the application of integrated control, which reduces to a minimum the disturbing effects of treatments on the environment. It proves to be more efficacious because it is better aimed, taking into consideration a series of ecological and ethological parameters in the ecosystem considered, and to avoid the use of persistent products with toxic, nondegradable constituents or polluting effects.

The aim has been pursued for several years in the member countries of the European Communities, with specific and thorough research which takes into consideration various agricultural crops in the framework of joint phytosanitary defence.

The risk of toxic chemical residues in crops intended mainly for human consumption, is always great and cannot continue or even be accepted.

For field vegetables the joint community research deals with precise aspects, such as ecological investigations and investigations into population dynamics, perfecting and standardising methods of observation and prevention of attacks, with the exact evaluation of the damage which has occurred, verifyng the applicability of biological control with the use of entomophages, pheromones and biopesticides, research for resistant cultivars, and the limitation of treatments with pesticides and herbicides.

This research is carried out with financial contributions from the European Communities of up to 50% of the total cost; in addition, work meetings, seminars, study visits, and the exchange of researchers are also carried out.

This meeting has been preceded by similar experts' meetings which proved particularly fruitful and useful for a better understanding between researchers and research of a common mode of operating and of a closer collaboration. They took place in Paris (F) in March 1979, Montfavet (F) in October 1980, Brussels (B) in December 1980, Dublin (IR) in September 1982, and again in Brussels in December 1983: they dealt with the deep investigation of phytosanitary aspects of one or more field vegetable crops.

This Rennes meeting was organised jointly by the CEC and the IOBC and is the result of the common will to operate together in a joint effort toward reaching as rapidly as possible the same objective.

The meeting has brought together eminent researchers and scientists, whom I have been privileged to welcome most cordially in the name of the Directorate-General Agriculture of the Commission of the European Communities. In addition to thanking them for their attendance, I give them my best wishes for fruitful work.

A particular and very cordial greeting goes to the President of the West Palaearctic Regional Section of the IOBC, Prof. C. Pelerents, who has honoured us with his presence, and to whom I should like to express my warmest appreciation and my gratitude for the active and close collaboration which we firmly intend to pursue in the future.

I thank most sincerely the INRA, Rennes Research Centre - Domaine de la Motte-au-Vicomte, for having made this meeting possible, and for the kind welcome and warm hospitality it has shown us. In particular I thank Dr. E. Brunel, in his capacity as representative of the local Secretariat, for his precious work in contributing to making the meeting as well run and efficient as possible.

The meeting will not only allow us to gain better knowledge of the phytosanitary problems which are being posed at present for field vegetables in the various countries involved, but to make a point of the advancement of research, and to discuss modern techniques and the development of alternative control methods. Above all it will permit us to point out a reality for which joint forces and strategies are necessary in a framework of technical and scientific collaboration, towards which we must tend with all our resources.

The setting up of a strong understanding and close cooperation on all levels between persons who work towards the same goals, is the most heartfelt wish I make to the participants. I add to this the hope that our work will soon the success that one expects in this very important sector of the efficacious protection and health of field vegetables.

Opening address by the International Organization for Biological and Integrated Control representative

C. Pelerents
President IOBC/WPRS

Ladies and Gentlemen,

The choice of Rennes for the first joint C.E.C./I.O.B.C. meeting is perhaps not strange when one remembers that the "Integrated control in Brassica crops" working group was formed here in June 1971, following the remarkable work performed by Prof. J. Missonnier on the ecology of Cabbage root fly.

The work of this working group was concentrated on this insect first. Close collaboration between researchers from different countries, after a certain time allowed the development of sampling methods and the establishment of tolerance thresholds.

Over the years other pests such as the Cabbage aphid and the Cabbage moth were considered and it was possible to develop an integrated control strategy.

With the setting up in 1978 by the Commission of the European Communities of imprtant research programmes centred on biological and integrated control, concerning in particular vegetables, it was possible to enlarge the research to other fields and to other vegetable crops.

Considering the complementary aspect of this researches, and to avoid duplication, a close and exemplary cooperation between the two organisations was carried out. It can now be realized in this joint experts' group meeting, which has a very full programme but which shows that the research has moved far beyond the Brassica crop.

In addition I am particularly happy to see among the participants experts in phytopathology, because it is the first time that the problem of disease in the integrated control context will be treated in detail. I should also like to stress the fact that this meeting has the task of drawing up proposals concerning the protocols of joint research.

In the name of my organization and in my own name, I should like to thank Mr. E. Brunel most sincerely for hving organised this meeting so efficiently and the National Institute of Agronomic Research for having offered their facilities for the meeting.

I am sure that the collaboration between the IOBC and the CEC will continue, and I should like to thank the CEC representative, Prof. R. Cavalloro, in advance. I hope that all the participants enjoy a fruitful meeting with profitable discussions.

Session 1
Brassica crops

Chairman: T.Coaker

Problems in monitoring cabbage root fly, *Delia radicum (brassicae)* by oviposition traps

H. den Ouden

Research Institute for Plant Protection, Wageningen, Netherlands

Summary

In 1984 and 1985, the use of Freuler's Swiss egg trap for recording oviposition of the cabbage root fly was compared in a preference and a non-preference situation. It is concluded that a non-preference situation gives a more reliable indication of early oviposition of D. radicum.

1. INTRODUCTION

Methods available for monitoring the cabbage root fly, Delia radicum (brassicae) (L.) are - besides the yellow water traps for the adults: 1) the collection and counting of eggs around the stem base of a certain number of plants or 2) the use of the Swiss egg trap according to Freuler and Fischer (1983). Their advice is to place e.g. ten specimens of this trap evenly spread over a cabbage field. They should be applied to the plants in a line in such a way that a choice is offered to ovipositing flies between plants with traps in certain lines on the one side and the plants without a trap in the same lines and those in the adjacent lines on the other. The question to be answered was how far this placement of the traps gives optimal qualitative and/or quantitative information about the presence of D. radicum.

2. MATERIALS AND METHODS

In 1984 an experiment was carried out in which a non-preference square field situation consisting of 49 cauliflower plants each provided with a Swiss trap was compared to a preference square field situation of 196 plants alternately provided with a trap (Figure 1a).

In 1985 a similar test was performed but now with one trap per eight plants as preference situation. Here 400 plants were placed in a rectangular field. The corresponding non-preference conditions were created by providing 8 x 7 plants each with a trap evenly placed in the field (Figure 1b).

In 1984, on May 22 and 28 and June 12, egg collection was done from soil around the stem base of the 98 plants without trap. Samples consisting of three spoons of soil were washed up in water and the floating eggs of D. radicum were sieved out and counted. At these dates the numbers of eggs obtained in this way could be compared with the captures from the traps. At June 5 and 18 only the eggs from the traps were counted for comparison of the preference and non-preference oviposition. The eggs around the stems were not collected at those dates but removed from the stem base for enabling a proper comparison of the weekly oviposition in traps and in the soil one week later. At June 25, however, no more samples were taken.

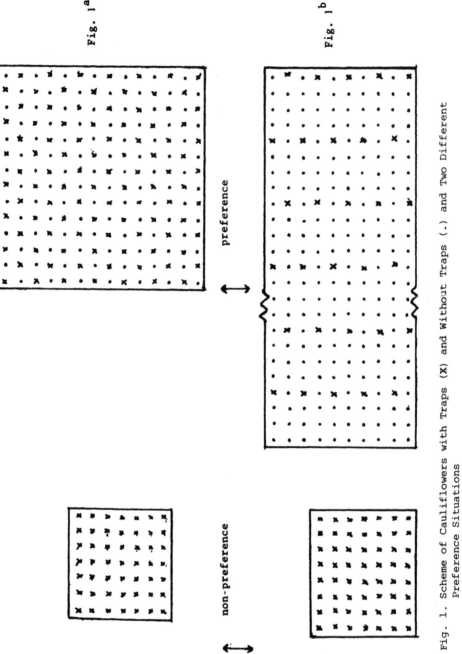

Fig. 1^a

Fig. 1^b

preference

non-preference

Fig. 1. Scheme of Cauliflowers with Traps (X) and Without Traps (.) and Two Different
Preference Situations

12

3. RESULTS (summarised in Table 1)

   Where a preference situation had been offered to the flies, significantly more eggs were collected from the soil around the stem base than from the traps. At the dates mentioned the ratios were: 10.7 (May 22), 2.1 (May 28) and 2.3 (June 12) at $P < 0.01$. Significantly more traps than stem base soil samples were found without eggs. The ratios were 9.6, 2.2 and 2.8 respectively for the dates above ($P < 0.01$).

TABLE 1. Ratios of Sampling Results in Different Situations in the Field
*, ** and *** means significance at $P < 0.05$, < 0.02 and < 0.01 respectively. - means not significant.

Dates	May 22 1984	May 28 1984	June 5 1984	June 12 1984	June 18 1984	May 28 1985
Figures compared:						
Stem base counts over trapping	10.7***	2.1***	.	2.3***	.	.
% samples without eggs trapping over stem base	9.6***	2.2***	.	2.8***	.	.
Trapping counts in non-pref. over pref.	2.94***	2.81***	1.49⁻	1.24⁻	1.02⁻	1.56**
% empty traps in non-pref. over pref.	1.56⁻	1.97*	0.81⁻	0.87⁻	0.88⁻	1.00⁻

   When the traps in a non-preference field were compared to those in a preference field at May 22, 2.29 and 0.78 eggs per trap were collected respectively. A χ^2-test showed a significant difference at $P < 0.01$. At May 28 these figures were 2.73 and 0.97 respectively, with the same significant difference. However, the samplings at June 5, 12 and 18 showed no significant differences between the preference and non-preference conditions.

   The reason behind this phenomenon might be found in the fact that in small plants the 'egg-laying mood' of the flies is relatively under developed when they arrive at the stem base. In the non-preference situation they move to another plant with trap and oviposit there adding to the egg number of plants with traps. In the preference situation the chance of moving to a plant with trap is 50% lower, so the egg numbers of trap-provided plants become smaller here. When plants become higher and the walk to the stem base is longer, the stress of moving to another plant decreases and the ratio between egg numbers in the preference and non-preference situation also decreases.

   The percentage of empty traps in the two situations did not differ significantly at May 22, 1984. At May 28, 1984 there was a weakly significant difference ($P < 0.05$).

   When only 12.5% of the plants were provided with traps, in the preference situation 16 eggs per trap were collected (n = 50) versus 25 per trap in the non-preference field at May 28, 1985 (n = 56). The difference is significant at $P < 0.02$. The traps with zero eggs in the preference field equalled those in the non-preference situation (4%). The numbers of traps with more than 30 eggs were 18 and 3, respectively.

4. DISCUSSION

The conclusion from these experiments is that quantitative monitoring in preference situations is an acceptable procedure only when sufficiently large numbers of eggs are deposited and tolerance levels are already exceeded.

The ratio of eggs per trap in non-preference and preference conditions seems to be dependent on density and time of the season. Another important factor might be plant height, although this actually should be proved in an experiment with small and large plants in the same plot.

The possibility to use egg number caught per trap as a fixed criterion for decision making in pest management therefore seems to be an illusion particularly because the standard errors are considerable. In the preference fields they amounted to 190% and 63% at May 22, 1984 and May 28, 1985, respectively. However, if groups of fourteen respectively eight traps placed in two lines are considered as one sample, the standard errors became 55% and 24%. In the wide placement of the traps the standard error is smaller but this is caused by the higher average number of eggs at the end of May 1985.

The distribution in 1984 was very skewed because of the large amount of zero counts. In such a situation a reliable figure for decision making on the base of a tolerance level can be obtained only when using a very large number of traps.

Freuler and Fischer (1983) mentioned a required number of 40 traps. This has been confirmed by Pelerents and van Keymeulen (1984). Taking into account the above-mentioned standard errors it might be more. However, for an extension service in a cabbage or typical cauliflower district, a special field for early warnings in the centre of the district might be an acceptable solution. In that case, a non-preference field would be preferable in order to obtain as many eggs in the traps as possible.

ACKNOWLEDGEMENTS

Thanks are due to Mr. J. van der Beek and Mr. S. Folkertsma for assistance during the experiments.

REFERENCES

(1) FREULER, J. and FISCHER, S., 1983. Le piège à oeufs, nouveau moyen de prévision d'attaque pour la mouche du chou, Delia radicum (brassicae) (L.) Revue Suisse Vitic. Arboric. Hortic. 15: 107-110.

(2) PELERENTS, C. and VAN KEYMEULEN, M., 1984. Methods for the monitoring and integrated control of cabbage root fly Delia radicum (brassicae) (L.). C.E.C. Programme on integrated and biological control. Final report 1979/1983: 273-285.

Use of egg-traps in a warning system against the cabbage root fly *(Delia radicum)*

B.Bromand

Institute of Pesticides, National Research Centre for Plant Protection, Lyngby, Denmark

Summary

A one year investigation with 10 egg-traps at each of 12 localities in Denmark has given the background for sending out warnings
for the first and second generation of the cabbage rootfly. The
traps were emptied twice a week by the growers and the results
reported by telephone to the Plant Protection Centre from where
the warnings were send out.

1. Introduction

In a cooperation project the relation between the catch of cabbage rootflies in yellow water traps and the number of cabbage
rootfly eggs in egg-traps was investigated in Denmark in 1984. A
very good agreement was achieved especially between the catch of
female flies and the number of eggs. The results will be presented by Martine van Keymeulen.
As a result of this investigation it was decided to make a warning system against the cabbage rootfly in 1985 based on egg-
traps placed on a number of localities spred all over the country.

2. Materials and Methods

Through agricultural advisors and research stations 12 localities
were chosen representing the main cabbage growing areas and
contact was made to the growers.

15

The system is based on the growers help and to each grower were
send 22 egg-traps, a letter with instructions and scemes to fill
in with the number of eggs counted in the traps.
10 traps were placed on cabbage plants a few rows from the border
of the field. 10 traps were used for changing and 2 as a reserve.
Each monday and thursday the traps were emptied and the results
telephoned to the Plant Protection Centre. The growers were asked
to report from May 9th to June 10th and again from June 20th to
July 25th, but some reported all the way through to September.
Most growers were very conscientious and there were very few
problems in getting the results in. In cases where the growers
reply were delayed for more than 2 days the grower was called on
the phone. Figure I. shows the different localities.

Figure I. Localities where egg-traps were placed

3. Results

The results appear in Tables I and II for first respectively
second generation. Tabel I shows show that the first generation
started about May 23rd, which was late due to the cold spring.
Normally first generation starts May 10th - 15th.
The localities I, II and 12 started 5 days later than most other
localities. I and II are located at the northern part ofJutland
and 12 at Falster, which in spring is greatly influenced by the
cold Baltic.
Table II shows a greater variation in the start of the egg-laying
of second generation as well as on the number of eggs in the
traps.
FigureII shows the average dayly number of eggs in 10 traps.
Second generation is by far the greatest. This is believed to be
dependent on the increasing area of oilseed rape. First generation multiply on oilseed rape, but when second generation is on
their wings the oilseed rape is unsuitable for egg-laying and the
flies seek out cabbage fields.
May 24th and July 10th warnings were send out to all agricultural
advisors and they were advised to set in control measures 5 days
later allowing for the eggs to develope.
Several growers reported that they had great advantage of visi-
ting the field so often keeping track of the activity of the cabbage rootfly.

4. Discussion

This investigation surely proved to be successfully already the
first year, but at the same time several problems still lacks an
answer.
Were the warnings send out at the right time and did we give the
right advise ? Obviously the dates of the warnings were as close
as they could be, but from the resultant attacks to judge the
control measures should have been applied at the time of the
warnings. This leds to the most important question about the
damage treshold. How should it be defined and at what level is
it ? Does it depend on previously applied control measures ?

Table I. Dayly number of cabbage rootfly eggs in 10 traps. 1. generation 1985

Locality	13.5	17.5	20.5	23.5	28.5	30.5	3.6	6.6	10.6	13.6	17.6	20.6	24.6	27.6	1.7	4.7
1. Hjørring	-	-	0	1	40	28	109	42	36	12	28	23	46	9	13	15
2. Øster Vrå	-	-	-	0	15	36	109	81	29	-	-	-	-	8	7	5
3. Gistrup	-	0	0	4	10	44	23	42	28	14	2	0	0	0	0	0
4. Spjald	0	0	1	13	41	-	21	24	8	-	-	-	-	-	1	0
5. Lystrup	-	0	0	10	15	15	43	13	9	-	-	-	-	-	1	0
6. Skast	0	0	0	2	1	-	3	5	0	3	0	0	0	0	0	0
7. Årslev	0	1	2	84	35	65	26	65	8	-	-	-	5	0	2	0
8. Skælskør	0	0	5	10	45	11	49	33	3	-	-	-	2	1	0	3
9. Jægerspris	-	-	0	59	131	23	22	34	4							
10. Lyngby V.	0	0	1	33	15	14	19	5	3	4	-	8	13	1	0	13
11. Lyngby Ø.	0	0	4	17	20	30	22	14	4	2	-	11	14	4	0	26
12. Nykøbing F.	-	-	0	0	6	4	9	3	0	-	-	-	0	5	0	0
Total	0	1	13	233	374	270	455	361	132	35	30	42	80	28	24	62
Average	0	0	1	19	31	27	38	30	11	7	10	8	11	3	2	6

Table II. Dayly number of cabbage rootfly eggs in 10 traps. 2. generation 1985

Locality	8.7	11.7	15.7	18.7	22.7	25.7	29.7	1.8	5.8	8.8	12.8	15.8	19.8	22.8	26.8	29.8
1. Hjørring	84	199	60	80	63	109	255	629	273	148	144					
2. Øster Vrå	0	8	5	73	78	244										
3. Gistrup																
4. Spjæld	12	4	-	28	25	81										
5. Lystrup																
6. Skast	0	23	104	144	-	136	186	51	47	97	150	62	55	13	3	2
7. Årslev	2	34	87	83	76	141	531	318	35	267	326					
8. Skælskør	2	7	84	47	72	190										
9. Jægerspris																
10. Lyngby V.	36	84	49	95	31	22	80	86	82	54	113	109	22	23	28	18
11. Lyngby Ø.	73	79	147	83	22	50	208	195	151	61	99	97	48	22	32	12
12. Nykøbing F.	1	2	22	58	24	29	68									
Total	210	440	558	691	391	1002	1328	1279	588	627	832	268	125	58	63	32
Average	23	49	70	77	49	111	221	256	118	125	166	89	42	19	21	11

Normal practice is to use carbofuran granulates at the time of transplanting, but this is not sufficient to control the cabbage rootfly especially not second generation.

In Denmark the following chemicals can be used for spraying: Shell Birlane 24 EC (chlorfenvinphos) with 4 l per ha, Basudin 25 Emulsion (diazinon) with 4 l per ha and Orthene 75 SP (acephat) with 10 l per ha.

Under severe attacks watering each plant with 0,1 l of diazinon has proved to be very successfull.

Conclusion

This first year of investigation has proved that it is possible to set up a warning system based on egg-traps and to give warning at the right time. Questions about the damage treshold and suitable control measures still have to be solved. These problems are subjects for future investigations.

Thermal requirements for cabbage root fly, *Delia radicum*, development

R.H.Collier
Department of Zoology and Comparative Physiology, University of Birmingham, UK
S.Finch
National Vegetable Research Station, Wellesbourne, Warwick, UK

Summary

Laboratory and field-cage experiments indicated that 6°C was the appropriate base temperature above which to accumulate day-degrees (D°) for forecasting the times of cabbage root fly activity under field conditions. The most appropriate way to record these accumulations was from integrating thermometers whose probes were placed 6 cm deep in the soil, the depth at which most of the insects' development occurs. A total of 580D° accumulated, above a base of 6°C, at a depth of 6 cm in the soil was required for the cabbage root fly to complete a generation. The number of D° required to complete a generation remained constant throughout all three annual generations of the fly.

1. INTRODUCTION

In common with other cold-blooded animals, the rates of most stages of insect development are directly proportional to temperature (1). Therefore, it is possible by determining the rates of pest development at a range of temperatures under controlled laboratory conditions to produce a simple model for forecasting temporal changes in field populations of the specific pest. Such models are usually based on a physiological time scale, that is measured in accumulations of some heat units such as 'day-degrees' (D°). Laboratory studies are essential for producing the basic framework of this type of model, as field conditions are far too variable for any type of controlled experiments.

The overall aim of the research is to develop a practical system for forecasting the times of appearance of the cabbage root fly so that control procedures against this pest can be made as effective as possible. This paper is restricted solely to describing the type of laboratory and field-cage studies carried out to produce the simple D° model used in the initial forecasts.

2. EXPERIMENTAL WORK

2.1. Determining the base temperature

Although the rate of insect development is proportional to temperature within the mid-temperature range of insect development, it deviates at low and high temperatures. Hence the plot of rate of insect development against temperature is sigmoid at the two temperature extremes (2). As a result, the usual way to determine the base temperature is to record the rate of insect development every 1-2°C over

21

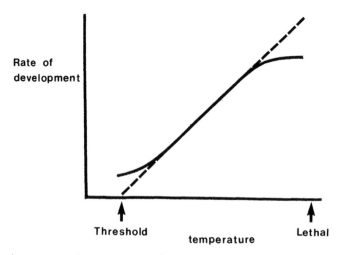

Fig. 1. <u>Method for determining the base temperature for development</u>

at least a 20°C range (Baker, 1980), to ensure that sufficient of the recorded points lie on the linear part of the curve. A regression is then fitted only to these points and extrapolated (Fig.1). The point where this line intercepts the abscissa is regarded as the theoretical base temperature but is not necessarily the developmental zero (1-3) (Fig. 1).

The base temperatures for development for each stage in the life-cycle of the cabbage root fly were estimated by rearing the various stages in a series of cooling incubators (Gallenkamp Ltd, London, England) maintained at constant temperatures of between 0° and 30°C. Linear regressions were then fitted to the data, in the manner described above, to produce the base temperatures shown in Table 1.

Table 1. <u>Time in days for the cabbage root fly to complete development of certain stages in its life-cycle at both 11° and 20°c and the appropriate base temperatures estimated from linear regressions.</u>

Developmental Stage	11°C	20°C	Base
	Development time in days		Temperature
Post-diapause	21	10.5	4.3
Pre-oviposition	15.5	5.5	6.6
Egg	9	3.5	6.4
Larval	48	17.5	6.8
Pupal	40	14	6.2

2.2 Calculation of D°

The number of day-degrees C (D°) was calculated using the method proposed by the British Meteorological Office (4) from the formula simplified for constant temperature studies, namely:-

D° = T50 at N° x (N°-base t°)

where N° is the temperature at which the insects were maintained, t is

22

the base temperature °C, and **T50** the time until 50% of the flies had emerged.

2.3. Confirmation of base temperature

To confirm that the estimated base temperature is the appropriate one, it is usual to calculate the number of day-degrees required for development from the 'predicted' base temperature and from bases that are one or two degrees higher and lower than the 'predicted' base. If the predicted base temperature is accurate, then the number of day-degrees required for subsequent development remains constant irrespective of developmental temperatures. Base temperatures that lie on either side of the true base produce curved rather than straight line relationships. Consequently, at low development temperatures, when the base chosen is too high the number of accumulated D° is underestimated and conversely when the base chosen is too low they are overestimated (5).

2.4. Choosing an appropriate base temperature

Although 4.3°C was the appropriate base temperature above which to accumulate day-degrees for post-diapause development (5), bases slightly in excess of 6°C were more appropriate for determining accurately the rates of development during pre-oviposition, egg hatch, larval and pupal development and fly emergence. Therefore, instead of complicating the system with a series of base temperatures, we have chosen the compromise of 6°C as our standard. This base was chosen because we are interested mainly in the numbers of D° separating the three annual generations. As post-diapause development occurs prior to emergence of the first generation of flies then its base of 4.3°C need not be considered in the prediction of fly activity during the summer months (5). At present, 6°C seems to be an appropriate standard base temperature above which to accumulate D° for the field forecasts.

2.5. D° to complete a generation

A schematic diagram of the number of D° required for development of the various stages that occur during a complete cabbage root fly generation are shown in Fig. 2.

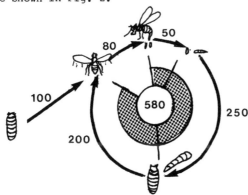

Fig. 2. Diagrammatic representation of life-cycle of cabbage root fly

The results show clearly that an accumulation of 580 D° above a base temperature of 6°C are required to complete a generation and that 500 of these D° are accumulated by root fly stages that live in the soil.

2.6. Recording accumulated D°

Although it is relatively simple to calculate the number of D° an insect receives under field conditions from the routine arithmetical calculations already mentioned, it is far easier to use some form of integrating thermometer. Once such thermometers are set to the base temperature, they display continuously on a visual display unit the numbers of D° accumulated during the test period.

In the field, the numbers of D° recorded on commercially-available integrating thermometers were similar to those calculated from both screen and soil temperatures obtained from an adjacent weather station. Integrating thermometers were therefore adopted for determining the temperature requirements of the cabbage root fly in the field.

Under field conditions, cabbage root fly larvae feed and pupate at depths of 2-10 cm. Although the D° accumulated by the integrating thermometers were slightly higher at a depth of 2 cm than at 10 cm, emergence was not greatly affected as flies emerged only one day (8D°) earlier from pupae 2 cm deep than from those 10 cm deep. Thus in practice it is not necessary for the temperature-sensing probes attached to the integrating thermometers to be placed at different soil depths during the year. Soil temperatures were recorded therefore at a depth of 6 cm to be representative of those experienced by the insects. In general, there was close agreement between laboratory measurements of rates of development and those recorded in the field.

2.7. D° changes with generation

To determine whether the D° requirements for a generation remained constant throughout a season (May-October), pots of radish, inoculated at twice weekly intervals with cabbage root fly eggs, were placed in a field cage and sampled at regular intervals. The pots were placed in field cages to prevent field flies laying in the test pots. All pots rested on the soil and were well-watered daily to keep the plants alive and to prevent the developing insects dying of desiccation. Temperature probes in the pots indicated that the same number of D° was required for each generation, irrespective of the week of inoculation.

Further studies also indicated that the date of inoculation, or oviposition, in one year did not affect the times the flies emerged in the following year, illustrating that winter effectively condenses overwintering pupae into just one population. The number of D° required for the cabbage root fly to complete a generation was approximately 580 D° above 6°C, measured at a soil depth of 6 cm, provided that the insects' development was not interrupted by either a summer aestivation or winter diapause at the pupal stage.

3. DISCUSSION

Under field conditions the accumulated D° required for each stage of cabbage root fly development were similar to those predicted from laboratory studies carried out at a range of constant temperatures. This suggests that the base temperature of 6° and the accumulated D° that we have estimated should provide a sound basis for developing a robust model for accurately forecasting cabbage root fly activity under field

conditions. There are severe problems facing such studies in both
laboratory and field tests. In the laboratory, the two major drawbacks,
to anyone contemplating a similar study on another pest species, are the
vast numbers of insects involved and the relatively large amount of space
and expensive equipment that is required. For example, the present study
occupied two rooms each with a floor space of 16 square metres, one to
house 12 cooling incubators and the other to rear the large numbers of
insects required for the tests. Without such facilities the time scale
involved would be prohibitive. Despite such facilities, problems arise
when working at temperatures close to the base temperature, as certain
tests may each last as long as six months. In such tests, large numbers
of insects have to be included from the start to ensure that sufficient
survive the extremely protracted developmental period to obtain a
reasonable estimate of the normal variations in development.

In the field, although comparable studies require less expensive
equipment it is essential to have a large area of space within either
large (3m x 6m x 2m high) or small (1m^3) insect-proof cages. Without
such cages it is impossible to exclude the wild population of flies from
the experimental treatments and this makes any meaningful interpretation
of the field results extremely difficult. In addition, field tests are
generally extensive and often involve weekly or bi-weekly tests to ensure
that the insects response does not change during the growing season.

Two additional difficulties in attempting to initially assess the
thermal requirements of the cabbage root fly under field conditions are
that the test insects may be induced into either aestivation or diapause,
the two phases of arrested development that occur in response to high and
low temperatures, respectively. Other problems are that in the field the
eggs and larvae may be eaten by predators and adult populations are often
unpredictably decimated by fungal pathogens. Furthermore, it is
practically impossible to destructively sample the roots of plants to
assess larval development without at the same time changing the overall
conditions that affect the remaining insects. Controlled laboratory and
field-cage tests appear therefore to be the most appropriate way of
obtaining data for the initial forecasting model.

REFERENCES

1. BAKER, C.R.B. (1980). Some problems in using meteorological data to
    forecast the timing of insect life cycles. EPPO Bull. **10**, 83-91.
2. SHARPE, P. J. H. and DEMICHELE, D.W. (1977). Reaction kinetics of
    poikilotherm development. J. theor. Biol. **64**, 649-670.
3. VAN KIRK, J.R. and ALINIAZEE, M.T. (1981). Determining
    low-temperature threshold for pupal development of the western
    cherry fruit fly for use in phenology models. Environ. Entomol.
    10, 968-971.
4. ANON. (1969). Tables for the evaluation of daily values of
    accumulated temperature above and below 42°F from daily values of
    maximum and minimum temperatures. Meteorological Office Leaflet
    no. **10**, 10pp.
5. COLLIER, R.H. and FINCH, S. (1985). Accumulated temperature for
    predicting the time of emergence in the spring of the cabbage
    root fly, Delia radicum (L.) (Diptera: Anthomyiidae). Bull. ent.
    Res. **75**, 395-404.

Factors regulating diapause and aestivation in the cabbage root fly, *Delia radicum*

R.H.Collier
Department of Zoology and Comparative Physiology, University of Birmingham, UK
S.Finch
National Vegetable Research Station, Wellesbourne, Warwick, UK

1. INTRODUCTION

Throughout most of the year, the rate of cabbage root fly development is directly proportional to temperature (1). There are two periods, however, one during winter and the other during hot summers when this relationship fails to remain operative because the insects enter a resting phase. The winter resting phase (diapause) is entered in response to a combination of low temperatures/shortening daylength and the summer resting phase (aestivation) in response to high temperatures during a critical stage in the insects' life-cycle. For any accurate forecast, therefore, it is essential to be able to predict when these resting phases will occcur and for how long they will persist.

This paper describes how, using an approach similar to that for estimating the thermal requirement for normal development (1), the factors regulating both diapause and aestivation have been quantified at Wellesbourne.

2. DIAPAUSE

2.1. Introduction

Cabbage root fly pupae overwinter in diapause and must receive a cold treatment to complete their diapause development. This must be followed by a period at warmer temperatures for their post-diapause development, which is complete when the flies emerge in the spring. To predict accurately the thermal requirements, or cumulative day-degrees (D°) above 6°C for emergence of flies in the spring, it is essential to determine when diapause has been completed and hence when to begin accumulating D°.

2.2. Experimental work

To determine the time that diapause is terminated under field conditions, cabbage root fly pupae were extracted from field soil every two weeks from November to April during the winters of 1979-82. The pupae were extracted from the soil by flotation (2). A minimum of 500 pupae was collected at each sampling. In general, about 1 kg of soil had to be processed to obtain each pupa.

Once diapause has been completed, most flies emerge within 14 days if pupae are maintained at 20° (3 & 4). This finding was used, therefore, as the criterion to determine the percentage of pupae that had completed diapause in each of the 2-weekly samples.

The results indicated that although there was a 2-3 week variation between the years, diapause had generally been completed in field populations by the beginning of March (Fig. 1.)

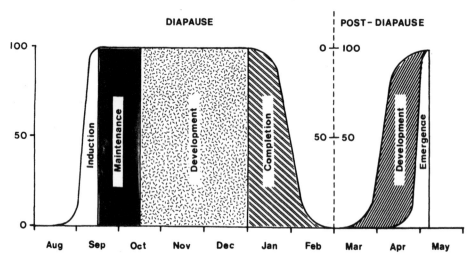

Fig. 1. <u>Schematic diagram of the phases entered by overwintering pupae of</u> <u>the cabbage root fly.</u>

In certain populations, cabbage root flies produced a bi-modal pattern of emergence in the spring (5) which could possibly represent the contributions from the second and third generation of the previous year to the overwintering population. To determine if this occurs, pots of radish were inoculated under field cage conditions from early August to late September to produce diapausing pupae comparable to those of the second and third generations. When samples of these pupae were brought into the laboratory at the times field pupae were sampled, it was evident that diapause had been terminated to the same extent in the pots as in the natural field pupae. Hence, the diapause resting phase during the winter effectively condensed individuals from the second and third generation into just one overwintering population (4).

Throughout these experiments, diapause was completed in more than 50% of the population usually by early-February (Fig. 1). February 1 was chosen, therefore, as the most appropriate date to start the D° accumulation for predicting peak emergence of the overwintering (first) generation of the cabbage root fly.

2.3. Discussion

Figure 1 shows schematically the stages entered by overwintering, diapausing cabbage root fly pupae. Diapause begins in mid-August when the first few larvae are subjected to conditions that induce the subsequent pupae into diapause. Warm autumn days then maintain (6) such pupae in diapause until temperatures are sufficiently low for diapause development to proceed. Diapause completion occurs in an increasing proportion of pupae from late December/early January onwards. Despite completing diapause so early in the winter, the early-pupating individuals cannot develop any further until temperatures become sufficiently warm in the spring for post-diapause development to start. This delay in development in the spring until the onset of warmer weather again slows down the early developers and helps to condense the two overwintering generations. Hence

28

the overwintering individuals usually emerge as an extremely discrete
population of flies within a very short period in the spring.

3. AESTIVATION

3.1. Introduction

Although the cabbage root fly is restricted to the temperate zone of
the holarctic region (latitudes 35-60'N) where mean air temperatures are
usually less than 20°, soil temperatures may rise to 30° or more in short
periods of hot weather. When such temperatures occur in mid-summer, all,
or a proportion, of the cabbage root fly population may enter a summer
resting phase (aestivation) in which development temporarily ceases. Once
the population, or a proportion of the population, has been induced into
aestivation, it becomes increasingly difficult to forecast accurately the
times when flies will be active during the second and third generations.

3.2. Experimental work

Large numbers of insects were reared continuously under laboratory
conditions at a wide range of both constant and fluctuating temperature
conditions. Many of the experiments were conducted in programmable
cooling incubators (7).
    The results showed clearly that cabbage root flies can only be
induced into aestivation shortly after pupariation, that is during the
early stages of pupation. If hot temperatures occur prior to this stage,
or shortly after it, they merely increase the overall rate at which the
insects develop.

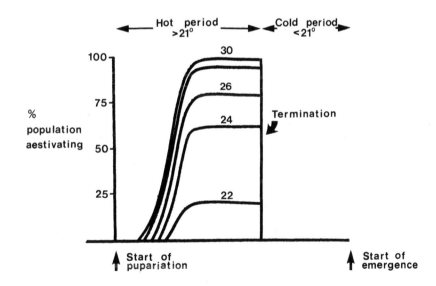

Fig. 2. The proportion of a cabbage root fly population induced into
         aestivation at five different development temperatures.

29

3.3. Discussion

Figure 2 shows schematically how aestivation can influence development of a cabbage root fly population. Once pupariation is completed, then the proportion of the population that enters aestivation depends upon the ambient temperature. At a constant temperature of 20°, few if any pupae enter aestivation, whereas at 28-30° practically 100% of the population aestivates. The pupae remain in this state of arrested development for as long as the high temperatures persist. Once temperatures again fall below 20°, however, development re-starts and most pupae soon pass out of the aestivation-sensitive phase. Although the delay in development of each individual insect lasts only as long as the hot period, the overall effect to any particular forecast depends ultimately on the proportion of the population susceptible to the aestivating conditions. In each cabbage root fly generation, the majority of eggs are laid during a concentrated 3 week period of oviposition and hence there is a period of a similar duration about a month later when variable proportions of the population can be induced into aestivation. Being able to forecast the susceptible proportion of the population when aestivating conditions occur is an essential refinement for improving the accuracy of the basic D° model.

4. CONCLUSIONS

The present studies on diapause and aestivation clearly indicate that the forecasting model based on D° accumulated above 6°C can now be improved by starting to accumulate the D° from 1 February. More importantly, however, the forecast can also be improved by incorporating corrections to account for when pupae in the field begin to enter diapause. Similarly, since the present results have defined 1) the conditions that induce aestivation, 2) how to determine the proportion of the population in the aestivation-sensitive phase and 3) the conditions required to terminate aestivation, it should be possible from local weather conditions to improve the accuracy of the forecast during seasons when aestivation becomes part of the normal life-cycle of the fly.

Although diapause and aestivation both induce cabbage root fly pupae into periods of arrested development, they have diametrically opposite effects on the activity patterns of the subsequent generations. Diapause condenses the spread of activity by not allowing any early-pupating insects (usually 2nd generation) to develop until late in the year when temperatures fall below 6°C. It also condenses the rate of development in the spring, as the individuals that complete diapause early during the winter period, have their development arrested until the appropriate warm conditions return in the spring. There are therefore two phases in which diapause condenses overwintering pupae and ensures that flies subsequently emerge as one discrete population in the spring. In contrast to diapause, aestivation generally spreads the activity of a particular generation (usually the mid-summer one) over an extended period, because unless conditions are exceptional (i.e. 100% aestivation (8)), only a proportion of the population is delayed by aestivation. Being able to predict the effects of both diapause and aestivation will help considerably to refine the current model for forecasting the times of cabbage root fly attacks.

REFERENCES

1. COLLIER, R.H. & FINCH, S. (1985). Thermal requirements for cabbage root fly, Delia radicum, development. (This Journal).
2. FINCH, S., SKINNER, G. and FREEMAN, G.H. (1978). Distribution and analysis of cabbage root fly pupal populations. Ann. appl. Biol. 88, 351-356.
3. COLLIER, R.H. & FINCH, 2. (1983). Completion of diapause in field populations of the cabbage root fly (Delia radicum). Ent. exp. & appl. 34, 186-192.
4. COLLIER, R.H. & FINCH, S. (1983). Effects of intensity and duration of low temperatures in regulating diapause development of the cabbage root fly (Delia radicum). Ent. exp & appl. 34:193-200.
5. FINCH, S. & COLLIER, R.H. (1983). Emergence of flies from overwintering populations of cabbage root fly pupae. Ecol. Entomol. 8:29-36.
6. MISSONIER, J. (1963). Etude ecologique du developpement nymphale de deux dipteres muscides phtyophages; Pegomyia betae Curtis et Chortophila brassicae Bouch. Annls. Epiphyt. 14:293-310.
7. FINCH, S. & COLLIER, R.H. (1985). Laboratory studies on aestivation in the cabbage root fly (Delia radicum). Ent. exp. & appl. 38:137-143.
8. FINCH, S. & COLLIER, R.H. (1985). Forecasting the times of attack of Delia radicum on brassica crops:- Major factors influencing the accuracy of the forecast and the effectiveness of control. (This Journal).

Emergence of cabbage root flies from puparia collected throughout northern Europe

S.Finch, B.Bromand, E.Brunel, M.Bues, R.H.Collier, R.Dunne, G.Foster, J.Freuler, M.Hommes, M.Van Keymeulen, D.J.Mowat, C.Pelerents, G.Skinner, E.Städler & J.Theunissen
(For full addresses see authors contributions; for countries see Table 1)

Summary

The patterns of cabbage root fly emergence throughout the mainland of northern Europe were equally as variable as those recorded in the United Kingdom. Only the populations from Denmark were as early-emerging as the Wellesbourne flies. In the remaining populations between 15% and 91% of the individuals emerged early. The greatest difference in a continental country occurred in Switzerland where 32% of the flies emerged early at Lausanne, compared to 91% at Wadenswil. These data indicate that forecasting the times of cabbage root fly attack in continental Europe from the accumulation of D° above a base temperature of 6°C will require the same type of extensive data that is currently being collected in the United Kingdom.

1. INTRODUCTION

In 1976, a population of cabbage root flies was discovered in Halsall, Lancashire that emerged extremely late in the year from the overwintering puparia (1). Prior to this finding, all populations studied (2,3,4 & 5) appeared to be of the early-emerging type similar to the population endemic to Wellesbourne (1). A small survey carried out in England and Wales from 1978-80 soon revealed, however, that late-emergence was not restricted to the Halsall site (1). The survey also showed that in general cabbage root fly populations could be grouped into early-, intermediate- or late-emerging types. This variation in the times of emergence of flies from the overwintering population obviously makes it much more difficult to predict accurately the times of cabbage root fly attacks under field conditions.

The present survey was carried out to determine whether the patterns of fly emergence throughout northern Europe were equally as variable as those recorded in the United Kingdom.

2. EXPERIMENTAL WORK

2.1. Materials and methods

Whenever possible, samples of 200 puparia were collected during the winter of 1982-83 and forwarded to Wellesbourne for this co-operative study. The two samples from France were forwrded during the winter of 1983-84.

The puparia were collected using a variety of methods. The commonest was to find a field where the crop had been harvested and where the

33

remaining roots showed signs of earlier, extensive larval damage. Once a suitable field was located, it was just a matter of carefully scraping the soil away from the vicinity of the damaged roots and hand-picking the puparia out of the disturbed soil.

All puparia were forwarded to Wellesbourne so that the patterns of fly emergence from the various populations could be studied under a standard set of environmental conditions. On the day puparia arrived, they were washed, counted, placed into jars of damp vermiculite and stored at 4°C until after the field population of flies had completed its emergence at Wellesbourne. Batches containing at least 50 pupae were then placed in a room at 20°C with a 16 h photoperiod and the numbers of flies and parasitoids that emerged were recorded. The numbers of the parasitoids of Aleochara bilineata Gyll., A. bipustulata (L.) (Staphylinidae), Phygadueon trichops Thoms. (Ichneumonidae) and Trybliographa rapae (Westw.) (Eucoilidae) that emerged from the various samples were forwarded to Dr B. Bromand for inclusion in his part of the co-operative study.

In these experiments, once the puparia had been placed at the post-diapause development temperature of 20°C in the controlled-environment room, any flies that emerged within the next fourteen days, were classified as 'early-emergers'.

2.2 Results

A summary of the patterns of emergence is shown in Table 1. The results have been expressed as the time for 50% (T50) of the flies to emerge, the figures being expressed as a percentage of the total number of flies that eventually emerged rather than as a percentage of the initial number of pupae. In this way, the influence of pupal deaths and the varying levels of parasitism are avoided. The spread of emergence from the various populations is also shown as T10-90.

More than 90% of the flies emerged within 14 days from the populations from Arhus, Skaelskør and Virumgaard (Denmark), Feltwell and Wellesbourne (England), Montfavet and Rennes (France) and from Wadenswil (Switzerland). These populations were considered therefore to be early-emerging. The remaining populations displayed a wide range of variable emergence patterns, and contained from 15% to 85% early emerging flies. The present populations did not fall into the three discrete groups that characterized the initial 13 populations from England and Wales (1) but instead were more or less spread equally over the whole range of possible combinations. Populations that were considered to have an intermediate-type of emergence were those in which from 30% to 85% of the population emerged within 14 days at 20°C. The times for 50% of the flies to emerge from such populations ranged from 8-9 days throughout the British Isles to 38 days in Belgium. Populations in which emergence was particularly late included those from Abergavenny (Wales), Gent (Belgium), Halsall (England) and Lausanne (Switzerland). Even after being subjected to cold temperatures for extended periods, the pupae from these populations still had the low rates of respiration that typify diapausing pupae. This finding suggests that these late-emerging populations either have two different phases of diapause or just one phase that is unusually protracted (6). The Feltwell population was included in the present survey because the sample of pupae collected in 1981 had been of an intermediate type (1). In the present survey, however, the population was more or less totally early-emerging, indicating that the characteristics of a population may change for reasons that are as yet unknown.

Table 1. Emergence of flies from puparia collected from several European
          sites during 1982-84.

Country	Locality	Collectors	%early emergers	T_{50}	T_{10-90}
Belgium	Gent	C. Pelerents	34	38	79
	"	M. Van Keymeulen			
Denmark	Arhus	B. Bromand	99	9	2
	Skaelskor	"	100	9	2
	Virumgaard	"	100	9	2
Eire	Dublin	R. Dunne	78	10	14
England	Feltwell	G. Skinner	91	8	8
	Halsall	R.H. Collier	15	33	43
	Wellesbourne	S. Finch	100	8	2
France	Montfavet	M. Bues	97	8	10
	Rennes	E. Brunel	94	8	6
Netherlands	Wageningen	J. Theunissen	58	11	24
Northern Ireland	Lisburn	D.J. Mowat	73	9	18
	"	S.J. Martin			
Scotland	Auchincruive	G.N. Foster	85	9	13
	"	D.J. Lamb			
Switzerland	Lausanne	J. Freuler	32	25	48
	Wadenswil	E. Stadler	91	9	8
Wales	Abergavenny	R.H. Collier	34	20	33
West Germany	Hürth Fichenich	M. Hommes	58	11	22

2.3. Discussion

The variations in the times of emergence of cabbage root flies from
these European populations of pupae indicated that there are two extreme
biotypes, one with early- (early May) and the other with late-emerging
(early July) flies. The absence of obvious steps in the emergence curves
of many of these populations indicated that intermediate-emerging
populations were not simple mixtures of early- and late-emerging biotypes,
but contained different proportions of individuals with a slightly
later-than-normal emergence characteristic (late May). Most of these
intermediate-populations were inclined more towards the early (normal)
than the late end of the emergence spectrum, and probably contained a high
proportion of hybrids.

The present results show clearly that the present method of
forecasting the times of cabbage root fly activity from accumulated
day-degrees may need to be refined considerably if it is to produce an
accurate forecast for more than a restricted number of localities. The
results show that, even within a region (e.g. Feltwell), the prediction

model may have to be adjusted to account for local differences in the rate of fly development affecting the times of appearance of flies in the spring. More importantly, control tactics may also need to be modified to account for the regional differences in the protracted spread of emergence in the spring (7) particularly if as a consequence of the spread of fly emergence the period of egg-laying in the spring also becomes extremely protracted.

3. ACKNOWLEDGEMENT

This work was carried out under MAFF licence No. PHF 428/92 issued under the Import and Export (Plant Health) and Plant Pests (Great Britain) Order 1980. We thank Miss V.L. Bowlzer for help in obtaining and renewing the above licence.

REFFERENCES

1. FINCH, S. and COLLIER, R.H. Emergence of flies from overwintering populations of cabbage root fly pupae. Ecol. Entomol. 8, 29-36.
2. COAKER, T.H. and WRIGHT, D.W. (1963). The influence of temperature on the emergence of the cabbage root fly, Erioischia brassicae (Bouche), from overwintering pupae. Ann. appl. Biol. 52, 337-343.
3. MISSONIER, J. (1963). Etude ecolgique du developpement nymphal de deux dipteres muscides phytophages; Pegomyia betae Curtis et Chortophila brassicae Bouche. Annls. Epiphyt. 14, 293-310.
4. HARRIS, C.R. and SVEC, H.J. (1966). Mass rearing of the cabbage maggot under controlled environmental conditions with obserations on the biology of cyclodiene - susceptible and resistant strains. J. econ. Ent. 59, 569-573.
5. MACLEOD, D.G.R. and DRISCOLL, G.R. (1967). Diapause in the cabbage maggot, Hyleya brassicae (Dipt. Anthomyiidae). Can. Ent. 99, 890-893.
6. COLLIER, R.H. and FINCH, S. (1985). Preliminary studies of the factors regulating late emergence in certain cabbage root fly, Delia radicum, populations. (This journal).
7. FINCH, S. and COLLIER, R.H. (1985). Forecasting the times of attack of Delia radicum on brassica crops:- Major factors influencing the accuracy of the forecast and the effectiveness of control. (This Journal).

Preliminary studies of the factors regulating late emergence in certain cabbage root fly, *Delia radicum*, populations

R.H.Collier
Department of Zoology and Comparative Physiology, University of Birmingham, UK
S.Finch
National Vegetable Research Station, Wellesbourne, Warwick, UK

Summary

Populations of late-emerging cabbage root flies occur quite commonly throughout northern Europe. Instead of emerging at the expected time during late-April and early-May, these populations usually emerge over a protracted period during June and July. It seems likely that late-emerging cabbage root flies have a different type of diapause to early-emerging flies. Investigations are now in progress to determine both the thermal requirements for development of these flies and the factors regulating their late-emergence. Additional studies are also being carried out to develop a suitable chemical method for the rapid identification of late-emerging populations.

1. INTRODUCTION

Accumulated temperatures (day-degrees) measured in the field have been used frequently to predict cabbage root fly activity, and although an accurate prediction has been produced at several individual sites, variation between sites has often been considerable (5, 11). A recent study to develop a robust forecasting system has shown that if day-degrees are measured in the appropriate micro-environment they can be used to predict accurately cabbage root fly activity (8). Finch and Collier (1985), however, have shown that there are differences in the patterns of emergence of cabbage root flies from overwintering pupae, in which the flies emerge both later, and over a more extended period, than was previously expected (6). Delayed emergence of this type occurs at certain sites in the United Kingdom and also in several regions of northern Europe (10). Such differences can occur within relatively short distances in a particular locality. There is also evidence of a gradual transition (clinal variation) from early- to late-emerging populations in one region of south-west Lancashire (9).

Late emerging cabbage root fly populations do not respond to temperature in a manner that can be predicted at present. For example, the standard cold treatment (6) does not terminate pupal diapause, and the post-winter rate of pupal development and subsequent fly emergence is not proportional to temperature. This suggests that there is possibly a further stage of diapause that requires a higher base temperature to enable the insects to complete their development.

The current forecasting model could probably be improved by determining the factors regulating these population differences. This applies particularly to such factors as the temperature requirements for late-emerging flies and the genetical basis producing such differences.

To develop an accurate forecasting model, it is important to understand the factors that regulate emergence in late-emerging populations. A study has therefore been started to subject both early- and late-emerging populations to a range of physiological, biochemical and genetical tests to determine how such populations differ from one another.

2. EXPERIMENTAL WORK

2.1 Emergence of populations of flies from the United Kingdom.

Diapause pupae sampled from several sites in the United Kingdom were taken to Wellesbourne, placed into Tygan (R) sleeves on the day after collection and then immediately reburied amongst a crop of swedes. The times for 50% of the flies to emerge from these pupae in the spring are shown in Table 1.

Table 1. Times and spread of emergence of flies from diapausing overwintering puparia collected from various regions of the British Isles and maintained under Wellesbourne field conditions.

Origin of puparia		Day 50% emergence reached	Spread (10-90%) of emergence in days
Country	Locality		
England	Wellesbourne	26 April	12
England	Feltwell	28 April	24
Wales	Abergavenny	9 June	66
Scotland	Auchincruive	29 April	40
Northern Ireland	Lisburn	15 May	51

All populations emerged later and had a greater (10-90%) spread of emergence than the Wellesbourne population. The results indicated that there is considerable variation between populations and that if a forecasting model is going to be of general application, then the differences must be quantified. The data indicated that relatively extreme differences in the times of emergence can occur between populations separated by only a few kilometres (9).

2.2. Timing of emergence

When diapause has been completed, most Wellesbourne flies emerge within 14 days if pupae are maintained at 20°C (2). This was therefore used as the criterion to determine the proportion of early-emerging flies in any given pupal population. Following this early emergence, there was then usually a period of about 14 days before the second batch of flies started to emerge. This second batch of flies continued to emerge over a period of 28 days, so that the last flies emerged approximately 8 weeks after being placed at the standard (20°C) post-diapause temperature. The second batch of flies were those described previously (2) as late-emerging. Similarly, once diapause development has been completed in the field and temperatures rise in the spring, the early-emerging flies begin their post-diapause development and subsequently emerge in mid-late April (3). This emergence is generally spread out over a 3-4 week period. In contrast to late-April, very few flies emerge in the field during May. However, flies begin to emerge again in early June and then continue to emerge over a protracted period lasting well into July.

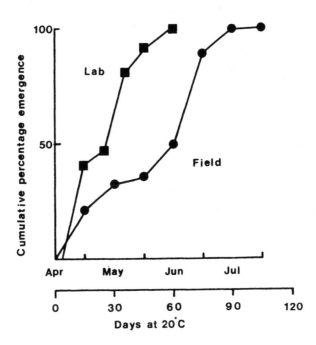

Fig. 1.   Cumulative emergence curves of flies from the populations of Abergavenny puparia maintained under both laboratory and field conditions.

Figure 1 shows the cumulative emergence curves for a cabbage root fly population collected from a crop of swedes at Abergavenny, Wales. Flies from one batch of these puparia were allowed to emerge in the laboratory at 20° soon after diapause had been completed whilst the others were reburied immediately into the plot of swedes described previously. The depressions in the cumulative emergence curves indicate that the population was clearly a mixture of early- and late-emerging flies. The emergence curves from the two batches of puparia are similar in shape but the timing of emergence of the flies differed considerably because of the different temperatures the puparia were exposed to under the laboratory and field conditions.

2.3. Is there an extended diapause?

When late-emerging cabbage root flies are maintained at 20°C after being subjected to low temperatures for 20 weeks, or more, they do not continue their development for at least a few more weeks. During this 'resting' phase their respiration rate is low and similar to that of early-emerging populations at the start of diapause. This suggests that late-emerging pupae remain in an extended diapause.

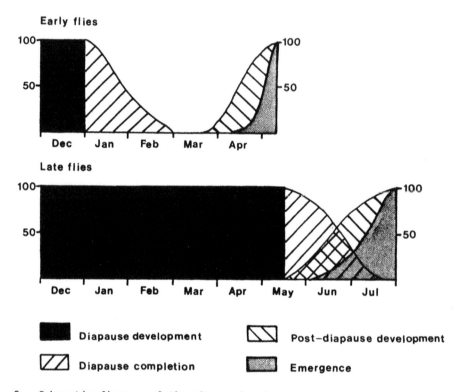

Fig. 2. <u>Schematic diagram of the phases involved prior to emergence of both early- and late-emerging populations of cabbage root fly.</u>

Figure 2 shows schematically the likely phases in diapause completion of both early- and late-emerging populations of cabbage root flies. Early-emerging populations complete diapause development over a period of several weeks during late December - February (1) but are prevented from starting their post-diapause development by low winter temperatures (3). Flies from these populations emerge in a highly synchronized manner when temperatures rise in the spring. Late-emerging populations appear to remain in diapause for several months longer, before diapause is spontaneously completed in the various individuals during either June or July. High mid-summer temperatures allow post-diapause development to be completed over an extremely short period and so there is no obvious time difference between diapause completion and subsequent fly emergence. This may help to explain why late flies emerge over a longer period (8 weeks) than early flies (3 weeks).

In early-emerging populations the early-developing individuals are faced with inappropriate conditions for post diapause development (temperatures too low) and so the late-winter/early-spring period effectively condenses the spread of both development and emergence of such populations. In late-emerging populations, however, there is no such constraint during the period of post-diapause development, because of the warm temperatures, and so the spread of emergence reflects the actual, rather than the condensed, spread of development of the flies within the particular population. Thus the timing and duration of emergence of late-emerging cabbage root flies depends largely upon the temperature requirements for diapause completion and is regulated far less by the requirements for post-diapause development.

2.4. The effects of temperature

Diapause pupae from several populations were exposed to low temperatures for periods of between 20 and 40 weeks and then placed at 20°C. The percentage of early-emerging flies did not increase following the 20 week exposure (2, 5). Similar experiments showed that the time for 50% of late flies to emerge became less as the period of exposure to cold conditions was extended. This change was extremely gradual, however, and accounted for only about a 1-2 week difference in the time of emergence under field conditions.

Preliminary studies to determine how temperature affects late-emerging individuals were made using the population of flies collected from Abergavenny. This population was chosen because it contained a mixture of both early- and late-emerging flies. Samples of pupae from this population were maintained at temperatures between 8 and 26°C in a series of cooling incubators. Late-emerging flies did not emerge either at temperatures below 8°C or above 23°C. The lack of response of such populations to temperatures below 8°C suggests they have a higher base temperature for diapause completion than early-emerging flies. The failure of late flies to emerge at temperatures above 23°C suggests that late-emerging pupae may complete diapause and then immediately enter aestivation. Earlier studies showed that early-emerging diapause pupae can be induced into aestivation provided the warm conditions occur before post-diapause development begins. In late-emerging flies, the optimum temperature for diapause completion appears to be about 15°C. Thermal requirements have been estimated for flies from the Abergavenny population maintained in the laboratory at 20°C and for those allowed to emerge normally in the field (Figure 1). Thermal requirements for 50% emergence of late flies were close to 500D° under both laboratory and field conditions but as Figure 1 shows, the spread of emergence was less easy to quantify.

3. DISCUSSION

3.1. General conclusions

The proportions of late- and early-emerging flies differ considerably between and within cabbage root fly populations (5, 6, 9).

Furthermore, in the field, late-emerging flies from the overwintering population can emerge at any time within a 4-8 week period during June and July. It is therefore extremely difficult to forecast cabbage root fly activities in localities where there are late-emerging flies. Such forecasts are made even more difficult by the facts that the proportion of late-emerging flies in a particular locality can change radically over a distance of a few kilometres and also from year to year at the same site. Such differences may occur as a result of changes in the cropping system or of the way the late-emerging trait is inherited. The overall aim of this study is to identify and quantify differences between cabbage root fly populations in the hope that the information will help to refine the present model for forecasting the times of attack by this pest.

3.2. Experimental approach

Studies are presently being carried out on cabbage root fly populations, from sites in south-west Lancashire and south Wales, that are known to have large proportions of late-emerging flies. Experiments are also being continued to estimate the thermal requirements of late-emerging flies so that these additional data can be incorporated into the forecasting model. In addition, a separate study is being continued to confirm that late-emerging individuals are in an extended phase of diapause. The chemical tests associated with this study are based mainly on gel-electrophoresis separation of the patterns of pupal proteins. Hopefully, the differences between pupae from early- and late-emerging populations in this study will allow populations of flies to be classified rapidly from pupal samples collected during the winter instead of having to wait until the flies eventually emerge in the field in the summer. Finally, a series of genetical tests are being carried out to unravel the factors regulating 1) the occurrence, 2) the inheritance and 3) the maintenance of the late-emerging trait within a specific locality.

3.3. Other effects of development

Although late emergence from overwintering pupae has a major effect on the activity of cabbage root flies in the United Kingdom, activity can also be affected as a result of the variation that occurs during other development stages. Several populations from Northern Europe have been studied to investigate differences that occur at other stages of development, such as during the period of non-diapause development, when the timing or spread of activity between populations can be variable. Variations in fly activity during this period could radically affect the shape of the peaks of fly activity during the summer months. As in other species, the combination of conditions of temperature and photoperiod that induce diapause may also vary with changes in latitude. Preliminary investigations have shown that the induction of diapause was high in certain populations subjected to temperatures that would normally prevent diapause occuring in the Wellesbourne population. Finally, the effect of temperature in regulating aestivation could vary between populations, so that the insects' response to particular weather conditions may vary from site to site. Information about these additional factors may also be required if the model is to be made sufficiently robust to make an accurate forecast for a wide range of sites.

REFERENCES

1. COLLIER, R.H. and FINCH, S. (1983). Completion of diapause in field populations of the cabbage root fly (Delia radicum). Ent. exp. & appl. **34**, 186-92.
2. COLLIER, R.H. and FINCH, S. (1983). Effects of intensity and duration of low temperatures in regulating diapause development of the cabbage root fly (Delia radicum). Ent. exp. & appl. **34**, 193-200.
3. COLLIER, R.H. and FINCH, S. (1985). Accumluted temperatures for predicting the time of emergence in the spring of the cabbage root fly, Delia radicum (L.) (Diptera: Anthomyiidae). Bull. ent. Res. **75**, 395-404.
4. COLLIER, R.H. and FINCH, S. (1985). Thermal requirements for cabbage root fly, Delia radicum, development. (This journal).
5. FINCH, S. (1976). Monitoring insect pests of cruciferous crops. Proceedings 1977 British Crop Protection Conference - Pests and Diseases. 219-226.
6. FINCH, S. and COLLIER, R.H. (1983). Emergence of flies from overwintering populations of cabbage root fly pupae. Ecol. Entomol. **8**, 29-36.
7. FINCH, S.and COLLIER, R.H. (1984). Development of methods fro monitoring and forecasting the incidence of Delia radicum (brassicae) populations on brassicas. In: Agriculture. C.E.C. Programme on Integrated and Biological Control. Final Report 1979/1983. (Eds Cavalloro, R. & Piavaux, A.). Luxembourg: Office from Official Publications of the European Communities, p. 287-302.
8. FINCH, S. and COLLIER, R.H. (1985a). Forecasting the times of attack of Delia radicum on brassica crops:- Major factors influencing the accuracy of the forecast and the effectiveness of the control. (This journal).
9. FINCH, S., COLLIER, R.H. and SKINNER, G. (1985). Local population differences in emergence of cabbage root flies from south-west Lancashire: implications for pest forecasting and population divergence. Ecol. Entomol. (In Press).
10. FINCH, S., BROMAND, B., BRUNEL, E., BUES, M., COLLIER, R.H., DUNNE, D., FOSTER, G., FREULER, J., HOMMES, M., VAN KEYMEULEN, M., MOWAT, D.J., PELERENTS, C., SKINNER, G., STADLER, E. and THEUNISSEN, J. Emergence of cabbage root flies from puparia collected throughout northern Europe. (this Journal).
11. NAIR, K. and McEWEN, F.L. (1975). Ecology of the cabbage maggot, Hylemya brassicae (Diptera: Anthomyiidae) in rutabaga in South Western Ontarion, with some observations on other root maggots. Can. Ent. **107**, 343-354.

Mortality of the immature stages of the cabbage root fly

S.Finch & G.Skinner
National Vegetable Research Station, Wellesbourne, Warwick, UK

Summary

Although there is 90% mortality during the life-cycle of the cabbage
root fly, only about 30% of this mortality now appears to occur at
the egg stage. Results indicate that most of the mortality among the
immature stages may be due to abiotic rather than biotic factors and
that the contribution of natural, or released populations, of
predatory ground beetles may be difficult to predict.

1. INTRODUCTION

Work at Wellesbourne in the late 1950's (1 & 2) and early 1960's
(3,4,5 & 6) showed that predators exerted an important check on the
populations of cabbage root flies, ground beetles alone killing more than
90% of the eggs and first-instar larvae produced during the first and
second generations. In recent years, however, averages of 150
third-instar larvae/plant have occasionally been recorded from certain
crops attacked by the first generation of flies, indicating that the
earlier estimates of predation may no longer be appropriate. Such
estimates were obtained when high residues of aldrin and dieldrin, the
commonest insecticides used at the time, were present in many soils.
These persistent chemicals killed a wide range of species other than the
target insects. Hence ground beetle predators surviving such hostile
environments may have relied heavily upon food, such as cabbage root fly
eggs, that arrived in the crop as a result of the immigration of pest
insect species. With other sources of food, e.g. worms, in short supply,
the ground beetles may have had little choice than to feed voraciously
upon cabbage root fly eggs. In contrast, it is possible that the use of
the non-persistent organophosphorus insecticides over the last 20 years
has allowed the systems to revert back to the balance they were in prior
to the dieldrin era. The present work, therefore, was designed to
investigate more closely the role played by predatory ground beetles in
regulating cabbage root fly mortality in the 1980's.

2. EXPERIMENTAL WORK

2.1 Materials and methods

In 1981, 1982 and 1983, 6m x 3m x 2m high insect-proof cages were
erected over areas of ground, half of which had been fumigated with methyl
bromide to kill beetle predators. Each cage was transplanted (60 cm
spacing between plants) with six rows of 12 cauliflower plants of either

cv. Finney's 110 for the first (spring) generation experiments, or cv. White Rock for the second (summer) generation experiments. Usually the plants within each cage were inoculated at random with one of nine (log scale) different numbers of eggs (treatments), with each treatment being replicated eight times. After 400 day-degrees above 6°C had elapsed (the heat units required to pupariation (7)), soil samples were taken from around the plant roots using a 15 cm diameter soil auger (8). The roots and shoots were weighed, and the pupae were washed out of the soil, counted and weighed. In addition to studying the effects of predators, the effects of irrigation were studied in 1982 and the timing of inoculation in 1983. The 1981-83 results were inconclusive. Therefore, at various times during 1985, 100 eggs were placed around 12 plants in each cage and the numbers of eggs eaten were assessed by sampling 3 plants a day for four days, the period normally taken for eggs to hatch under field conditions (9).

2.2 Effects of Predation

In all years the fumigation treatment was highly effective as no weeds emerged within the cages covering the fumigated soil. When pitfall traps were used in 1983, however, small numbers of beetles were caught within the sterilized cages. Throughout the three years of the experiment, cabbage root fly mortality was highly variable. For example, in 1981 root fly mortality was much lower in the cages covering the fumigated soil, whereas the reverse trend occurred in 1982. It would appear, therefore, that although beetle predators have been shown to eat large numbers of the immature stages of the cabbage root fly in the past, we have not been able to confirm this finding under 1980 conditions.

Similarly when root fly eggs were placed around caged plants and then recovered on each of four consecutive days, there was only an accumulative loss of 9±2, 13±3, 28±4 and 26±4 of the eggs over the 4-day period. Hence, egg mortality prior to hatching now appears to be more in the order of 20-30% rather than the 95% recorded earlier.

2.3 Effects of irrigation

When 10mm of irrigation was applied once a week, more root fly larvae survived, whereas when the irrigation was applied twice weekly fewer survived, possibly because pathogenic bacteria became more effective in the permanently damp soils. As expected, all irrigated plants grew larger and had lower R.D.I's than their non-irrigated counterparts, confirming the earlier findings of Coaker (10) that additional water helps brassica plants to outgrow, or tolerate, damage.

2.4 Effects of number of eggs inoculated

In all years, overall mortality increased with increasing egg density. Mortality rose from 50% when 10 eggs were inoculated/plant to 85% when 100 eggs were inoculated/plant. Overall mortality in the field at Wellesbourne, where up to 200 eggs are laid/plant during each of the first two generations, usually averages 85-95%. In the present experiments, mean pupal weight declined from 14mg when 10 eggs were inoculated/plant to 9mg when 500 eggs were inoculated/plant. In contrast, total pupal biomass rose from 70mg at 10 eggs/plant to 205mg at 300 eggs/plant and then levelled out. R.D.I. (11) varied between 70% - 92% and the damage threshold of 30% (11) was exceeded when 30-100 eggs were inoculated/plant.

2.5 Instant versus protracted inoculations

In the spring experiments in 1983 when the weather was abnormally cold and wet, egg mortality was extremely high and ranged from 90% at low to 98% at high egg densities regardless of whether the inoculation were instant or protracted.

In the summer experiments, when the weather was warm and water was adequate for larval establishment, eggs inoculated gradually over a protracted 5 week period suffered mortalities ranging from 85-95% at the different densities. When the same numbers of eggs were inoculated in one batch, mortality was lower and ranged from only 72% at 10 eggs/plant to 85% at 500 eggs/plant. The establishment of newly-hatched larvae appeared to be inhibited by larvae already on the roots indicating that a density-dependent factor limited the size of the population when weather factors were not limiting.

3. CONCLUSIONS

Our experiments failed to demonstrate that predatory ground beetles consistently kill a large percentage of the immature stages of the cabbage root fly. They did, however, demonstrate that weather is important in limiting cabbage root fly populations. Also, in the absence of other limiting factors, competition between larvae appears to limit their numbers in a density-dependent way. The factors that regulate this mechanism are unknown. It is possible 1) that older larvae are carnivorous, 2) that they produce a repellent to prevent newly-hatched larvae from establishing, or 3) that the environment caused by their feeding is hostile to young larvae.

The results of this study raise doubts about whether experiments to try and improve the effectiveness of predatory ground beetles should be attempted. If abiotic factors are responsible for a high percentage of the overall mortality, then there may be no advantage from releasing more efficient predators if they only kill those life-stages of the cabbage root fly that would normally by killed by abiotic factors.

REFERENCES

This list of papers is restricted to work carried out at Wellesbourne. Similar papers reporting high root fly mortality from predatory ground beetles were produced during the same period in several European countries. We have restricted our comparisons to the Wellesbourne data as both the earlier 1959-63 and the recent 1980-85 experiments were carried out in the same fields.

1. HUGHES, R.D. (1959). The natural mortality of _Erioischia brassica_ (Bouche) (Dipt., Anothomyiidae.) _J. exptl. Biol._ 37, 218-223.
2. HUGHES, R.D. and SALTER, D.D. (1959). Natural mortality of _Erioischia brassicae_ (Bouche) (Dipt., Anthomyiidae) during the immature stages of the first generation. _J. anim. Ecol._ 28, 231-241.

3. HUGHES, R.D. and MITCHELL, B. (1960). The natural mortality of
   Erioischia brassicae (Bouche) (Dipt., Anthomyiidae): life tables
   and their interpretation. *J. anim. Ecol.* **28**, 343-357.
4. WRIGHT, D.W., HUGHES, R.D. and WORRALL, J. (1960). The effect of
   certain predators on the number of cabbage root fly (*Erioischia
   brassicae* (Bouche)) and the subsequent damage caused by the
   pest. *Ann.appl. Biol.* **48**, 756-763.
5. COAKER, T.H. and WILLIAMS, D.A. (1963). The importance of some
   Carabiadae and Staphylinidae as predators of the cabbage root
   fly, *Erioischia brassicae* (Bouche). *Ent. exp. et appl.* **6**,
   156-164.
6. MITCHELL, B. (1963). Ecology of two carabid beetles, *Bembidion
   lampros* (Herbst.) and *Trechus quadristriatus* (Schrank). *J.
   Anim. Ecol.* **32**, 377-392.
7. COLLIER, ROSEMARY, H. and FINCH, S. (1985). Thermal requirements for
   cabbage root fly, *Delia radicum*, development. (This Journal).
8. FINCH, S. and SKINNER, G. (1980). Mortality of overwintering pupae
   of the cabbage root fly (*Delia brassica*). *J. appl. Ecol.* **17**,
   657-665.
9. COAKER, T.H. and FINCH, S. (197). The cabbage root fly *Erioischia
   brassicae* (Bouche). *Rep. natn. Veg. Res. Stn for 1970*, 23-42.
10. COAKER, T.H. (1965). The effect of irrigation on the yield of
    cauliflower and cabbage crops damaged by the cabbage root fly.
    Pl. Path., **14**, 75-82.
11. ROLFE, S.W.R. (1969). Co-ordinated insecticide evaluation for
    cabbage root fly control. *Proc. 5th Br. Insect. Fungic. Conf.*,
    1, 238-243.

Forecasting the times of attack of *Delia radicum* on brassica crops: Major factors influencing the accuracy of the forecast and the effectiveness of control

S.Finch
National Vegetable Research Station, Wellesbourne, Warwick, UK
R.H.Collier
Department of Zoology and Comparative Physiology, University of Birmingham, UK

Summary

Under field conditions cabbage root fly activity is influenced considerably by a) the times the flies emerge from the overwintering puparia, b) by factors inducing aestivation and, c) by the influence of day-length on the induction of diapause. By making detailed studies of these factors, we are now able to forecast more acurately both seasonal and local variations in the timing of cabbage root fly infestations.

Despite such forecasts, however, certain populations are still difficult to control, mainly as a result of 1) having to apply insecticide to established crops during mid-summer, 2) from certain farmers growing brassica crops in inappropriate areas, and 3) from a second species, the turnip fly (Delia floralis), also damaging plants in certain localities. The influence of these factors on the current strategies for pest control are discussed.

1. INTRODUCTION

In the British Isles, brassica crops (e.g. cauliflowers and cabbages) grown over a short season are protected effectively from the cabbage root fly by insecticides applied to the soil at sowing or transplanting. For crops grown over a long season (e.g. Brussels sprouts and swedes), however, the soil-applied insecticides do not persist sufficiently long to prevent the plant roots being damaged by larvae of the later fly generations. Generally such damage passes unnoticed on crops such as Brussels sprouts since, once the plants are established, they can tolerate the damage without it having a measurable effect on yield. It is only when late damage directly affects the part of the crop used for human consumption, such as the 'roots' of swedes, that it becomes troublesome by reducing the quality rather than the quantity (yield) of the crop. To prevent this occurring, many growers now make additional mid-season applications of insecticide as a form of 'cosmetic' protection.

A recent agronomic change in the cultivation of swedes has also exacerbated the problem of protecting this crop. Prior to 1975, farmers drilled swede crops as near as possible to 1 June. In recent years, however, the seeding date has been changed from 1 June to 1 April to obtain the higher yields possible from the 2-month longer growing season. This means that present swede crops can be attacked by as many as three generations of cabbage root fly in one year, with the insecticide applied at drilling being effective against only the first.

1.1. Aim of the research

The aim of this research is to develop a practical system for forecasting the times of appearance of the cabbage root fly so that control procedures against this pest can be made as effective as possible.

Methods for forecasting the times of insect attack from accumulated temperatures (1, 2) are generally produced by determining the rate of development of the insect under a range of laboratory conditions. Such information is then used to predict when the various stages of the insect should be active in the field. At the same time, the accuracy of such predictions is assessed by monitoring cabbage root fly activity under field conditions, using methods developed specifically for sampling the various stages of the insect (3). By combining these two approaches, it is possible to determine those parts of the insects' life-cycle that need to be studied in greater detail to produce a forecast sufficiently accurate for use by growers.

This paper identifies three areas in the life-cycle of the cabbage root fly that are major factors in determining the accuracy of the present forecasts.

2. EXPERIMENTAL WORK

2.1. Factors influencing accuracy of forecast

Apart from the over-riding influence of temperature, the combined influences of, a) emergence, b) aestivation and c) photoperiod largely regulate the times cabbage root fly populations reach their peak activity within a particular year. Simple descriptions of these factors and the way they express their effects are as follows:-

a) Emergence - The time and rate at which the local population of flies emerges from the overwintering puparia is reflected in the timing of subsequent generations.

b) Aestivation - Is the phase that occurs when insects pupating during the summer months are prevented from developing further by high temperatures.

c) Photoperiod - This has an effect when the daylength falls below a critical level, which in central England occurs on about 31 July.

It is easier to separate the interractions between these factors by considering them, as shown below, in the reverse order to which they become effective in the field.

2.2. Influence of photoperiod

Cabbage root fly development is directly related to temperature so that the insect requires less calendar time (though not physiological time) to complete a generation in warmer than in cooler years. This is an important factor in forecasting the times of attack in central and southern England, as it is only in warm years, commonly referred to as 'early seasons', that the insect manages to complete three full generations (Solid line - Fig. 1A). For the population to produce a sizeable (damaging) third generation, many eggs of the mid-summer (second) generation have to be laid before the daylength falls below a critical level, on or around 31 July. Any eggs laid after this time produce offspring that go directly into the overwintering generation (diapause) instead of emerging as adults for the third generation. Hence, in cool wet years (e.g. 1985), the second generation becomes so delayed that for all intents and purposes there is no third generation (Dashed line - Fig. 1A).

2.3. Influence of aestivation

Cabbage root fly can be induced into aestivation only during the early part of the pupal stage (4). Since this stage often occurs over a 2-3 week period there are obviously many years when a proportion of the population aestivates. The current model contains parameters that estimate the proportion of the second (summer) generation affected by aestivation. Fig. 1B illustrates two examples, one in which all of the population is affected by aestivation (Solid line - Fig. 1B) and the other in which only 50% of the population is affected (Dashed lines - Fig. 1B).

When all of the population is affected, and aestivating conditions persist for several weeks during late-June and early-July, it is possible to delay the second generation so that most eggs are laid after 31 July. As a result, there is again no appreciable third generation (Solid line -Fig. 1B).

When only 50% of the population is induced into aestivation, there can be different effects on the separate halves of the population. For example, if half of the population has passed the critical stage by the time high temperatures occur, then the high temperatures merely speed up the overall rate of development of those individuals so that they emerge correspondingly earlier. The other half of the population, however, remains in aestivation for as long as the high temperatures persist. Hence, this can separate considerably the two halves of the generation (Dashed line - Fig. 1B - Gen. 2a & 2b). When 'fractionation' of a generation occurs in this manner, the relevant question to be answered is "Are there sufficient insects present during any one period to merit the application of an insecticide?" This question obviously become more difficult to answer when the aestivating conditions are highly variable and split the second generation into several rather than only two peaks. Fig. 1B (Dashed line) also illustrates that the half of the second generation induced into aestivation may again emerge too late for its offspring to contribute to the third generation.

2.4. Emergence of flies in the spring

For any accurate forecast, it is essential to categorize the rate at which flies emerge from overwintering puparia in the spring, as this is reflected strongly in the peaks of activity of subsequent generations. Nevertheless, provided the time for 50% emergence is used as the criterion for all populations, it is just as easy to forecast accurately the timing of late-emerging populations (Dashed lines - Fig. 1C) as early-emerging ones (Solid D. radicum lines - Fig. 1C) (5). For an accurate forecast, however, the truly late-emerging individuals within a locality have to be considered as a separate population.

3. DISCUSSION

An important fact to remember is that even if the activity of a particular populations can be forecast accurately, it does not necessarily

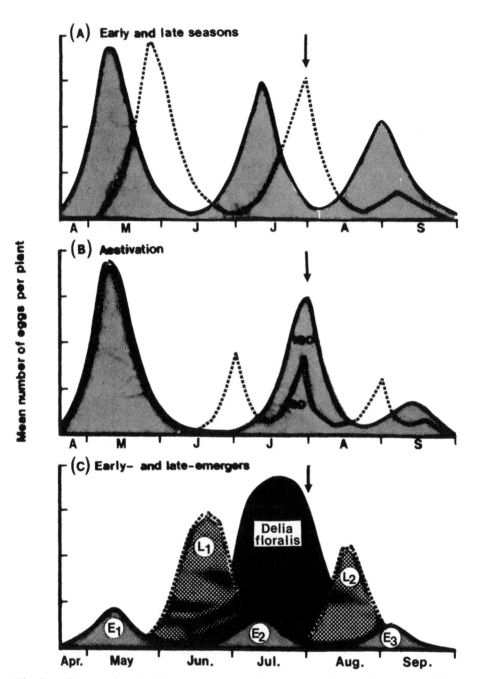

Fig 1.   Changes in the times of peak cabbage root fly activity resulting from (A) early (solid line) and late (dashed line) seasons, (B) aestivation in 100% (solid line) and 50% (dashed line) of the population and, (C) early- (solid line) and late-emerging (dashed-line) populations.   = 31 July; E_{1-3} = early- and L_{1-2} = late-emergers.

mean that such a population will be easy to control. This is particularly true during the mid-summer period when, because many crops are well-established, the insecticide has to be applied to the crop foliage or to the surface of the soil where it is much less effective than when incorporated into the soil (6). Therefore, unless mid-summer applications are made at critically-determined times they often fail to kill sufficient insects to produce an adequate control.

Controlling this pest can therefore become a problem a) when weather factors delay pest attacks; b) when insecticides have to be applied to well-established crops; and c) when for social reasons brassica crops have to be grown in inappropriate localities.

3.1. Effects of weather

In cool seasons when emergence of the first generation is relatively delayed, larvae are rarely subjected to the high levels of insecticides placed prophylactically around brassica plants shortly after transplanting. As a result, a higher proportion of their offspring survive to enter the second generation. In late seasons these insects lay most of their eggs during late July and early August, when egg deposition is often extremely heavy. If August is then cool and wet, few larvae die from desiccation (7). Under such conditions, even well-established (40 cm high) Brussels sprouts and cauliflower plants close to harvest can be subjected to severe water stress, though not usually killed, by the severity of the attack. Fig. 1 also shows that periods of both low (Dashed line - Fig. 1A) and high (Solid line - Fig. 1B) temperatures can delay emergence of the second generation by a comparable amount.

3.2. Practical control problems

Problems in controlling this pest arise whenever individuals emerge late. Such problems arise not from poor forecasting but rather from the lack of any appropriate control measure. The current trend of selecting relatively non-persistent insecticides is obviously an inappropriate strategy for populations having protracted periods of emergence. Furthermore, most of the insecticide applied into the soil at transplanting has normally lost its effectiveness by the time late-emerging populations become active. In addition, flies that emerge from overwintering puparia are larger than the offspring from the first generation and, since they lay more eggs (8), they produce a correspondingly heavier infestation. In midsummer, the infestations from the late-emerging populations are usually much heavier than those from the early-emerging populations and so increase the difficulty of post-planting protection.

3.3. The ultimate challenge

The most difficult place to control the cabbage root fly in England occurs at Halsall, Lancashire, where the root fly population emerges over a protracted period, often as separate groups of early- and late-emerging individuals. Insecticides applied early in the season in such localities influence only the proportion of the overwintering population (generally less than 20% of the total - Solid line D. radicum - Fig. 1C) that is early-emerging. From early June onwards, however, there is a more or less constant insect pressure (Fig. 1C), which has to be controlled by surface or foliar applications of insecticides. As the first generation of the late-emerging part (approximately 80% - Dashed line - Fig. 1C) of the population has overwintered, its flies lay large numbers of eggs. To make matters worse, the turnip root fly (Delia floralis Fall.) also occurs in this region and is active during mid-summer (Fig. 1C). This co-existence between the two species maintains a high level of root fly pressure throughout a major part of the brassica growing season. The case is not always as simple as the one illustrated in Fig. 1C, however, as D. floralis is more discriminating than D. radicum and often restricts itself to a particular brassica crop in a locality, while at the same time avoiding others. The factor regulating this choice appears to be host-plant age rather than host-plant type.

Controlling cabbage root flies in the Halsall area of south-west Lancashire is exacerbated by social problems. In this region, some farmers have traditionally relied on small, sequentially-planted strips of brassica as their major source of income. This was satisfactory when the organochlorine compounds were available. It is far from satisfactory at present, however, as the Halsall locality is mainly an area of peat soil, for which there is currently no approved organophosphorus treatment effective against this pest. If brassica crops have to be grown in such regions, a practice that should be discouraged whenever possible, then a prophylactic rather than a timed application of insecticide, albeit relatively ineffective, appears to be the only control strategy available for the present.

REFERENCES

1.   ECKENRODE, C.J. and CHAPMAN, R.K. (1972). Seasonal adult cabbage maggot populations in the field in relation to thermal-unit accumulations. Ann. Ent. Soc. Am. 65, 151-156.
2.   COLLIER, ROSEMARY H. and FINCH, S. (1985). Accumulated temperatures for predicting the time of emergence in the spring of the cabbage root fly, Delia radicum (L.)(Diptera: Anthomyiidae). Bull. ent Res. 75. 395-404.
3.   FINCH, S. (1980). Pest assessment - as it relates to Hylemya brassicae populations. Integrated Control in Brassica Crops. IOBC/WPRS Bulletin 1980/111/1, pp 1-9.
4.   FINCH, S. and COLLIER, ROSEMARY H. (1985). Laboratory studies on aestivation in the cabbage root fly (Delia radicum). Entomol. exp. appl. 38, 137-143.
5.   FINCH, S. and COLLIER, ROSEMARY H. (1983). Emergence of flies from overwintering populations of cabbage root fly pupae. Ecol. Entomol. 8, 29-36.
6.   SUETT, D.L. (1977). Influence of soil tilth and method of application of chlorfenvinphos on its persistence and distribution in soil and uptake by cauliflowers. Med. Fac. Landbouww. Rijksuniv. Gent 42, 1779-1788.
7.   FINCH, S. and SKINNER, G. (1983). Cabbage root fly. Mortality in the soil. Rep. natn. Veg. Res. Stn. 1982, 34-35.
8.   FINCH, S. and COAKER, T.H. (1969). A method for the continuous rearing of the cabbage root fly Erioischia brassica (Bouche) and some observations on its biology. Bull. Ent. Res. 58, 619-627.

Experiences with action thresholds in cabbage crops in the Federal Republic of Germany and first results of a collaborative field trial

M.Hommes
Federal Biological Research Centre for Agriculture and Forestry, Institute for Plant Protection in Horticultural Crops, Braunschweig, FR Germany

Summary

Two types of action thresholds were tested on white cabbage, red cabbage and savoy in the field. Thresholds of type A based on the infestation level of the pests by counting the insects while thresholds of type B based on the percentage of infested plants by caterpillars or aphids. The number of sprays and the amount of insecticides was reduced obviously by spraying according to the thresholds A and B. First results of a collaborative field trial of the IOBC/WPRS-working group of integrate control on field vegetables, in which simple action thresholds of type B were tested are presented.

1. Introduction

By the use of action thresholds basing on visual field checks the amount of insecticides can be reduced without increasing risk of yield losses. In comparison with other common monitoring techniques action thresholds have the advantage tha they base on the real attack of the field which have to be treated. Therefore the number of sprays could be reduced on a minimum. Time and money saved for pesticides and sprays have been brought up against the efforts for checking the plants in the field and for computing the actual infestation level. The sampling method should be as simple as possible and shouldn't take much time (4). By the use of monitoring techniques the period for field checking could be reduced and also the time for checking with the use of a variable-intensity sampling technique (2).

In the last years two different types of spraying thresholds (A and B) were tested on several cabbage crops at the institute in Hürth-Fischenich. For threshold A the number of larvae of the different pest groups and the percentage of plants infested with small and large aphid colonies have to be determinated. Thresholds have been developed at the institute in Hürth-Fischenich for various cole crops (1). Method B is very simple and based on the percentage of plants infested by caterpillars or by the cabbage aphid. The number of larvae or aphids per plant is not relevant. In the Netherlands threshold of this type have been developed for Brussels sprouts, white and red cabbage (3,5). Saving time and requiring no special

Table 1. Action thresholds A   tested in 1984 and 1985

CATERPILLARS/100 PLANTS

- 200 larvae (up to 15 mm long) from those pests which lay
      their eggs as egg batches, like Mamestra brassicae

- 100 larvae (up to 15 mm long) from those pests which lay
      their eggs singly, like Pieris rapae

- 50 larvae of Plutella xylostella

- 25 larvae of all species with more than 15 mm long
----
susceptibility factor: 0,5 for cauliflower, broccoli, chinese
and savoy cabbage;   1,5 for colrabi, Brussels sprouts and red
cabbage

APHIDS

- 40 % infested plants with small colonies (up to 100
     aphids per plant)

- 10 % infested plants with large colonies (more than 100
     aphids per plant)
----
susceptibility factor: 0,5 for cauliflower, broccoli, chinese
and savoy cabbage

---

knowledge to distinguish the different pest species are the
advantages of method B. The disadvantage of this method is tha
control measures can not be tuned to a special pest attack.

2. Materials and Methods

     All tests were carried out at the experimental field of
the institute for plant protection in vegetable crops in Hürth
Fischenich. Common used varieties of savoy cv. Ice Queen, red
cabbage cv. Marner Lagerrot (1984) and Marner September (1985)
and white cabbage cv. Dauerweiß (1984) and Marner Frico (1985)
were transplanted (spacing 50 cm x 50 cm) in beds consisting o
4 rows of 40 m length. Each bed was divided into 5 plots of 8
length. A split plot design with 5 replications was used.
Sprays were done by a special plot spraying machine applying
600 litres/ha. Treatments consisted of an untreated check, a
fortnightly routine spraying with a tankmixture of deltamethri
(5 g AI/ha) and pirimicarb (150 g AI/ha) for the control of
caterpillars and aphids and sprays according to the thresholds
A and B. In experiment 2, where planting sets of white cabbage
were tested, routine spraying and spraying by threshold B were

Table 2. Action thresholds B

year tested	crop		% plants infested with caterpillars	aphids
1984	savoy		10	5
	red cabbage		30	10
1985	white and red	/first two checks	10	10
	cabbage	/ later checks	5	5

Table 3. Number of sprays and insecticide applications of the treatments in experiment 1

		TREATMENT	
number of	routine	thresholds A	thresholds B
SAVOY			
sprays altogether	5	4	5
deltamethrin application	5	2	4
pirimiphos application	5	3	3
RED CABBAGE			
sprays altogether	6	3	4
deltamethrin application	6	1	1
pirimiphos application	6	2	3

not done. 5 (in experiment 1 and 2) and 10 plants (in experiment 3) respectively were sampled per plot in a weekly or two weeks interval. The average value over all 5 plots per treatment was used for determinating the infestation level of the pests, comparing them with the thresholds and making the spraying decisions. In table 1 and 2 the values of the two types of thresholds are listed. According to the infestation level of caterpillars and aphids control sprays with deltamethrin or pirimicarb (same amounts as in the routine plots), were done singly or as a tank mixture. At harvest 20 plants per plot were evaluated and the number of not marketable heads accordin to the damage by caterpillars or aphids was registrated.

Data were submitted to analysis of variance and means wer separated by Duncan's multiple range test.

3. Results and Discussion

The results presented in table 4 do not show a significan difference in the efficiency between a regular insectide appli cation and spraying on the basis of thresholds A and B, al though the number of sprays and the amount of insecticides wer reduced obviously (table 3). On red cabbage more sprays and

Table 4. Effect of action thresholds on the yield of savoy
and red cabbage in 1984 (experiment 1)

	SAVOY		RED CABBAGE	
	number (%) of not marketable cabbage heads damaged by			
treatment	caterp.	aphids	caterp.	aphids
untreated	53 a	17 a	19 a	15 a
routine sprays	2 b	7 ab	0 b	3 b
sprays according thresh. A	2 b	8 ab	2 b	2 b
sprays according thresh. B	0 b	3 b	2 b	5 b

Means with the same letter are not significantly different

Table 5. Number of sprays and insecticide applications of the
treatments in experiment 2

	TREATMENT				
number of	routine (fictive) SET 1 - 4	thresholds A			
		SET 1	SET 2	SET 3	SET 4
sprays altogether	6	3	4	6	3
deltamethrin application	6	1	1	1	1
pirimicarb application	6	2	3	6	2

insecticides could be economized as on savoy. For example only
one application of deltamethrin was necessary for controling
caterpillars on red cabbage whereas 2 and 4 sprays respectivel
were necessary on savoy. The number of sprays and the amount o
insecticides was generally reduced more by the use of thresholds A than by B. These results agree with further experiments
done in Hürth-Fischenich (1).
    In experiment 2 thresholds A were tested on 4 different
planting sets of white cabbage (table 5 and 6). In all sets
there was only one insecticide application for controling the
caterpillars. One single spray was sufficient with the exception of set 1. In set 1 a rather high percentage of not marketable cabbage heads was recorded. this was caused by a high
infestation level of Pieris rapae larvae just below the thresh
old value, wherefore a further control treatment was not done.
Controlling the cabbage aphid in this experiment was unsatisfactory. Several sprays (2-6) have been done to control this
pest, because the infestation level transgressed the threshold
values time after time. The percentage of not marketable heads
concerning cabbage aphid infestation at harvest in the untreated plot was so low that no spraying has been necessary.
This result could be explain by the following factors. During
the summer 1984 a high infestation pressure by alatae cabbage
aphids was observed, but by reason of several heavy rainfalls
no large population could be build up. Or it may be a quality

Table 6. Effect of spraying thresholds on the yield of 4 different planting sets of white cabbage in 1984 (Experiment 2)

set (time of planting)	UNTREATED number (%) of not marketable cabbage heads damaged by		THRESHOLDS A	
	caterp.	aphids	caterp.	aphids
1 13.6.	24	2	10	0
2 11.7.	30	4	4	0
3 18.7.	38	2	0	0
4 1.8.	40	2	0	0

Table 7. Number of sprays and insecticide applications of the treatments in experiment 3

number of	TREATMENTS		
	routine	thresholds A	thresholds B
WHITE CABBAGE			
sprays altogether	6	2	3
deltamethrin application	6	1	3
pirimiphos application	6	1	3
RED CABBAGE			
sprays at all	6	2	3
deltamethrin application	6	0	2
pirimiphos application	6	2	3

of the cultivar, where the aphids were generally observed only on the wrapper leaves.

In experiment 3 a regular insectide application was again compared with sprays based on thresholds of type A and B. This trial is part of a collaborative field experiment of the IOBC-working group, where thresholds of type B were tested on white cabbage in several european countries. The aim of this project which is coordinated by the author and which started in 1985 is to evaluate thresholds under various conditions and to use the results for developing relevant thresholds in a shorter time. At the institute at Hürth-Fischenich the investigations were also done on red cabbage. The results of the experiment are presented in table 7 and 8. For white cabbage they are simiular to those of experiment 1. The use of action threshold clearly reduced the number of sprays and the amount of insecti cides. The retrenchments were higher with threshold A than wit B. On red cabbage thresholds B worked well too, while thresholds A were too high for caterpillars. The threshold value was not reached and therefore no insecticide was applicated for controling these pests. In this year Pieris rapae was the

Table 8. Effect of action thresholds on the yield of white and
red cabbage in 1985 (experiment 3)

treatment	WHITE number (%) of not marketable cabbage heads damaged by		RED CABBAGE	
	caterp.	aphids	caterp.	aphids
untreated	62 a	22 a	33 a	35 a
routine sprays	0 b	3 b	2 b	1 b
sprays according thresh. A	8 b	4 b	37 a	4 b
sprays according thresh. B	3 b	2 b	3 b	3 b

Means with the same letter are not significantly different

predominate species. A rather high number of larvae could be
observed during the whole growing period, whereas larvae of
Mamestra brassicae which was generally the most abundant
species were recorded on a very low level.
    Summarizing the results of the 3 experiments it is pointe
out that sprays based on thresholds could reduce both the
number of sprays and the amount of insecticide obviously. The
economizing effect was higher with threshold A than with B. Bu
therefore the risk was higher by thresholds A. In two cases,
concerning controlling of caterpillars the use of the action
thresholds have failed. This can be attributed to a high attac
by larvae of Pieris rapae. Therefore the threshold values for
this pest have to be decreased.

REFERENCES

1. HOMMES, M. (1984): Integrierte Bekämpfung von Raupen und
    Blattläusen im Kohlanbau. Mitt. Biol. Bundesanst. Land-
    Forstwirtsch. Berlin-Dahlem H. 218, 85-107.
2. HOY, C. W.; JENNISON, G.; SHELTON, A. B. and ANDALORO, J.
    T. (1983): Variable-intensity sampling: a new technique fo
    decision making in cabbage pest management. J. Econ.
    Entomol. 76, 139-143.
3. THEUNISSEN, J. and DEN OUDEN, H. (1983): Praktijkbemonster
    ing van plagen in vollegrondsgroenten 3. Geleide bestrijd-
    ing in rode kool. Gewasbescherming 14, 119-124.
4. THEUNISSEN, J. (1984): Supervised control in cabbage crops
    theory and practice. Mitt. Biol. Bundesanst. Land- Forst-
    wirtsch. Berlin-Dahlem H. 218, 76-84.
5. THEUNISSEN, J. (1985): Tolerance levels for supervised
    control of insect pests in Brussels sprouts and white
    cabbage. Z. ang. Ent. 100, 84-87.

Oilseed rape crops as a source of cabbage root fly infestations for cruciferous vegetable crops

G.Skinner & S.Finch

National Vegetable Research Station, Wellesbourne, Warwick, UK

Summary

Cabbage root fly populations in oilseed rape crops grown on light, friable soils now warrant close monitoring as some populations are reaching levels threatening the rape itself. Such high populations could, if allowed to establish, be disastrous for more susceptible vegetable crops such as cauliflowers which under enhanced insect pressure may no longer be protected adequately by available control measures.

1. INTRODUCTION

The area of oilseed rape in England and Wales has increased about ten-fold since 1974, from about 24,500 ha to about 260,000 ha. Meanwhile the area of cruciferous vegetable (mainly brassica) crops, totalling about 55,000 ha, has remained unchanged. Consequently, brassica crops are now overshadowed five-fold by the area of land planted with oilseed rape (1).

In 1974, a farming contractor introduced approximately 100 ha of oilseed rape into the crop rotations around Kineton, Warwickshire, approximately 8 km south of Wellesbourne. As a similar area was drilled again in the autumn of 1975, cabbage root fly pupae were sampled at the end of the spring generation of flies in 1976 to determine how many insects this crop had supported. Forty 15 cm diameter soil cores (2), each containing 3-5 plant roots, were collected from each of four approximately 10 ha fields. The numbers of cabbage root fly pupae extracted by flotation from the soil cores were 0, 0, 0.1 and 4 m^2 of crop. When similar samples were taken in 1984, 0, 26, 31 and 63 pupae/m^2 were recovered from four comparable fields. In the field that did not contain pupae, the soil was heavy clay.

Although all rape crops are not infested, the populations in those that are infested are a cause for concern as the cabbage root fly populations in some fields near Wellesbourne have apparently increased 24-fold in the last 8 years(1). This paper describes quantitative evidence of the numbers of cabbage root fly that can be found in crops not treated with insecticide to control this pest.

2. EXPERIMENTAL WORK

Winter oilseed rape crops, sown to achieve a theoretical stand of 100-120 plants/m^2, were heavily damaged by cabbage root fly at Long Itchington (Warwickshire) and at Newton on Trent (Lincolnshire) in September 1984. In many areas of these fields, the plant stand had been reduced to less than 40 plants/m^2. A stand of at least 60 plants/m^2 is considered necessary to produce an economic yield.

Table 1. Distribution of cabbage root fly damage across oilseed rape fields during the winter of 1984-5.

Location, Field size, cultivar	Crop No.	Direction of transect	RDI of 10 plants per 20m intervals across field					Mean RDI	Pupae /plant
Long Itchington 7.2 ha, Jet Neuf	1	N-S	52	41	22	57	27		
			61	39	49	52	44	44	1.0
		E-W	42	40	82	54	17		
			28	32	42	87	47	47	1.2
Long Itchington 9.2 ha, Bienvenu	2	N-S	17	8	34	20	30		
			17	0	32	20	30	19	0.4
		E-W	24	12	0	23	10		
			11	24	15	0	13	13	0.3
Eathorpe 10.0 ha, Rafal	3	N-S	39	38	20	34	18		
			20	25	20	25	13	25	0.7
		E-W	37	44	32	12	17		
			32	20	29	47	44	31	0.4
Thelsford 8.0 ha, Bienvenu	4	E-W	8	0	0	15	0		
			2	5	5	0	5	4	0.1
		E-W	2	5	5	0	4		
			8	0	10	0	6	4	0.1
Newton-on-Trent 5.0 ha, Rafal	5	E-W	55	60	39	32	44		
			45	48	46	36	18	42	1.1
		E-W	35	40	39	60	27		
			56	51	47	30	59	44	1.1

Both the heavily-damaged fields (Crop Nos 1 & 5) and three others in Warwickshire (Table 1) were sampled in October 1984 to determine the root damage index (RDI) (3) and the numbers of pupae overwintering in the various crops. Samples were collected from transects roughly at right angles to each other in N-S and E-W directions. Every 20m a 2m cane was randomly dropped onto the crop and 10 plants touching the cane were dug up to assess the RDI. At the same time, one 15 cm diameter soil core was taken from the neighbouring locality. Each soil sample contained the roots of 2-3 oilseed rape plants. In October 1984, even the heavily-damaged fields with RDI's higher than 40 produced only 1 pupa per plant, or 40/m^2. Nevertheless, the data showed clearly that RDI was directly related to the numbers of pupae (Fig. 1).

Therefore taking five 10-plant samples at random across a field seems an appropriate way of estimating the overwintering root fly population.

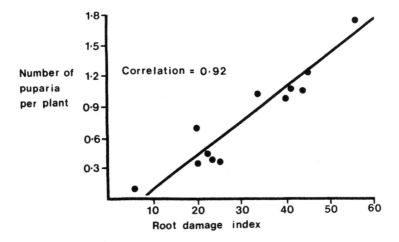

Fig. 1.   Relationship between numbers of cabbage root fly puparia
          overwintering in the soil and the root damage index in
          oilseed rape crops drilled into light soils during mid-late
          August 1984.

Table 2.   Mean numbers of cabbage root fly pupae recovered from around
           roots of oilseed rape plants during crop harvest in August 1985.

Warwickshire Locations	Cultivar	Site No	Pupae /sample	Pupae /plant	Pupae* /m^2	Mean Wt pupae (mg)
Long Itchington	Jet Neuf	1	61	18	1080	14
Long Itchington	Bienvenu	2	29	9	540	12
Eathorpe	Rafal	3	8	3	180	10
Thelsford	Bienvenu	4	1	1	9	9
Long Itchington	-	6	17	6	360	13
Hunningham	-	7	5	2	120	12
Wellesbourne	-	8	2	1	42	8
Harbury	-	9	13	5	300	13

*Based on areas of the crop where the plant stand was 60/m^2

    In the spring of 1985 a further 12 fields of oilseed rape were
sampled on the heavier clay soils to the West of Wellesbourne. In all
of these fields the RDI never exceeded 6%.
    In August 1985, four of the five fields on the light soils were
again sampled for pupae shortly after the oilseed had been harvested.
The numbers of pupae had by then risen to as high as 18/plant on the
most heavily damaged crop, the equivalent of 1080 pupae/m^2 (Table 2).
In contrast, there was still less than 1 pupa/plant under the crops on
the clay soils.

3. DISCUSSION

The influx of oilseed rape into the Wellesbourne area has increased considerably the problem of controlling cabbage root fly infestations in vegetable brassica crops. Previously, the root fly populations overwintered in the soil beneath crops of Brussels sprouts and swedes, the remains of which were ploughed-in during spring cultivation. In late-spring, the emerging adults were forced therefore to disperse and locate newly-drilled/transplanted brassica crops. As plants in such crops were generally small, those not protected by insecticide were ofted killed even by modest infestations. Despite the apparent susceptibility of these early crops, little damage occurred provided the plants had been treated with an appropriate insecticide. Spring treatments are relatively successful even against high levels of infestation, mainly because the insecticide can be incorporated into the soil at drilling/transplanting, the method of application that is by far the most effective (4).

The introduction of oilseed rape crops into crop rotation systems has produced a considerable increase in the infestation levels of this fly. At the start of the expansion of the oilseed acreage, most crops were drilled in the spring. Hence when such crops became established they competed with the vegetable brassicas for the first generation of flies and consequently helped pest control by effectively lowering the numbers of insects infesting each plant. The system has now changed dramatically and practically all of the 260,000 ha is drilled in the autumn. Hence the third generation flies are now provided with a choice of either small succulent oilseed rape plants or tough/senescing Brussels sprouts and swedes. It is not really a choice, and as a result high root fly infestations occur in oilseed rape crops in certain localities. Table 1 shows clearly that, on soils favourable to root fly, damage is spread evenly throughout an infested crop. In spring, emerging flies are no longer forced to disperse and many remain to lay around the bases of the actively-elongating oilseed rape plants. The success of the insect on such crops is evidenced by the numbers of puparia recovered in the autumn (Table 2). By autumn, however, the rape has been harvested and the large numbers of flies that emerge must again locate suitable host crops. When a new oilseed rape crop has been sown, then the cycle can continue. When no rape is available and large populations emerge adjacent to vegetable-growing areas, the results can be devastating. In both 1984 and 1985 at Wellesbourne, the mid-summer generation of flies was much larger than the spring generation and above-average late-season damage was clearly visible on vegetable crops during September.

The increased levels of attack on vegetable crops arose mainly because the new oilseed rape crops were not sown until mid-September and so the large numbers of flies that emerged from the harvested oilseed rape crops dispersed into the vegetable crops. Even if there are few insects/plant in an oilseed rape crop it may, because of its higher (80 plants/m^2) plant density, still provide a considerable threat to the majority of vegetable crops (maximum density 17 plants/m^2 for cabbage). Furthermore, as the oilseed rape acreage is five times that of vegetables, this could also potentially increase overall root fly problems. The reason why the country-wide increase in the population levels of this pest has not produced damage of catastrophic proportions is somewhat fortuitous. It has failed to occur mainly because 1) oilseed rape is grown as a break-crop largely in cereal areas that are free of vegetable crops, 2) oilseed rape has not become a dominant crop in the traditional vegetable-growing areas, as the vegetables themselves are the preferred 'break' crop and, 3) in certain localities where there are heavy clay soils on which both Brussels sprouts crops and oilseed rape are grown in close proximity, the soil conditions do not favour root fly survival. Table 2 illustrates the latter point particularly well. For example, from the heavy clay soils of fields 4 and 8 (Table 2) less than one insect was recovered per plant and the resulting puparia were small (8-9mg), a general indication that larval development was far from optimal.

Within the present cultivars there does not appear to be any inherent resistance to root fly damage, as the three commonly-grown cultivars Jet Neuf, Bienvenu and Rafal were all heavily damaged (Table 1). Although the Jet Neuf crop in the Long Itchington area (Field 1) appeared to be more susceptible than the Bienvenu crop (Field 2) (Table 2), this arose merely because the Jet Neuf crop was sown earlier in the season. It appears that, once a population of flies builds up in a crop in the autumn (Table 1), a large proportion of the offspring remain in the crop until it is harvested. Hence, the two-fold difference in damage recorded between Field 1 and 2 in November 1984 (Table 1) was reflected as a two-fold difference in the numbers of puparia recovered from the soil in August 1985 (Table 2).

The simple method of taking five 10-plant samples to estimate RDI during October/December of the year in which the crop is sown seems an extremely useful technique for forecasting whether a particular oilseed rape crop poses a threat at harvest to nearby vegetable crops. The nearer the two crops are, the greater the potential threat.

Whatever happens, it is unwise to allow cabbage root fly populations to increase in areas growing vegetable brassicas. In such localities, crop production is dependent upon applying insecticides that kill between 70-90% of the root fly larvae (5). For present infestations, the current insecticides provide effective control. If infestations increases, however, as a consequence of oilseed rape, present insecticides may no longer prove effective (1).

In localities such as Long Itchington and Newton on Trent, cabbage root fly populations are now so high that insecticides have to be applied at drilling to produce a reasonable plant stand even within the oilseed rape crops. Contemplation of wide-scale treatment of oilseed rape crops was feared since such a recommendation would increase the overall insecticide pressure and with it the chance of producing fly strains resistant to current insecticides. Surveys have shown, however, that many rape crops are drilled either in September when few flies remain to infest the crop or into clay soils that are unfavourable for survival of cabbage root fly larvae. Oilseed rape crops that pose most threat are those drilled early (mid-late August) into light, friable soils. Whenever possible, such crops should be drilled late or insecticide should be applied at drilling, if there is a history of heavy cabbage root fly damage in the locality. Both the Long Itchington and the Newton on Trent sites are in areas free of vegetable crops and so presumably their initial infestations arose from allotments, back gardens and wild host plants (6). Irrespective of the origin of the flies, they multiplied in the absence of insecticide into extremely high populations within three years of introducing oilseed rape into crop rotations on these light, friable soils.

REFERENCES

1. WHEATLEY, G.A. and FINCH, S. (1984). Effects of oilseed rape on the status of insect pests of vegetable brassicas. Proc. 1984 Br. crop Prot. Conf. - Pests and Diseases 8B-3, 807-814.

2. FINCH, S. and SKINNER, G. (1980). Mortality of overwintering pupae of the cabbage root fly (Delia brassicae). J. appl. Ecol 17, 657-665.

3. ROLFE, S.W.R. (1969). Co-ordinated insecticide evaluation for cabbage root fly control. Proc. 5th Br. Insect, Fungic. Conf., 1, 238-243.

4. SUETT, D.L. (1977). Influence of soil tilth and method of application of chlorfenvinphos on its persistence and distribution in soil and uptake by cauliflowers. Med. Fac. Landbouww. Rijksuniv. Gent 42, 1779-1788.

5. WHEATLEY, G.A. (1978). Biological activity of soil-applied pesticides in relation to method of application. Proc. 1977 Br. Crop. Prot. Conf. - Pests and Diseases 3, 973-984.

6. FINCH, S. and ACKLEY, C.M. (1977). Cultivated and wild host plants supporting populations of the cabbage root fly. Ann. appl. Biol. 85, 13-22.

Analysis of cabbage root fly population measurements and the derivation of control decisions

M.Van Keymeulen
IWONL, Centrum voor geïntegreerde bestrijding van insekten, Faculteit van de Landbouwwetenschappen, R.U.G., Gent, Belgium
C.Pelerents
Laboratorium voor Dierkunde, Faculteit van de Landbouwwetenschappen, R.U.G., Gent, Belgium

Summary

During 1983 and 1984 the efficiency and selectivity of Freulers egg-traps for the cabbage root fly were tested on nine cauliflower fields in Flanders. The egg- and adult captures were compared. To determine the optimal site and the number of egg-traps needed, the spatial distributions of eggs and adults were studied. For both insect stages, the relationship between variance and mean obeys Taylor's power law. With females captures only a very rough estimation of the eggs layed can be made. For fields with less than 15,000 cauliflowers (half a hectare) the beginning of oviposition can be determined by placing 10 egg-traps in one corner and by preference adjacent to a road.

Résumé

Lutte contre la mouche du chou, basée sur l'analyse de la population

L'efficacité et la sélectivité des pièges à oeufs (type Freuler) ont été testées en 1983 et 1984 dans neuf champs de choux-fleurs en Flandres Orientale et Occidentale. Les captures des femelles ont été comparées à celles des oeufs. Afin de déterminer l'emplacement et le nombre optimum des pièges, les distributions spatiales des oeufs et des adultes ont été étudiées. Pour ces deux stades, la relation entre la variance et la moyenne correspond à la loi de Taylor (fonction puissance). Seule une estimation très approximative des oeufs pondus peut se faire à partir du nombre des femelles capturées. Pour des champs d'environ 15.000 choux-fleurs (1/2 ha) le début de l'oviposition peut être déterminé en plaçant 10 pièges à oeufs à moins de 3 mètres du bord et de préférence dans un coin du champ.

1. Introduction

The cabbage root fly, *Delia radicum* L., inflicts important economic damage to cruciferous crops in Belgium. Especially cauliflowers are susceptible to *D. radicum* attack. The economic injury level at which no damage occurs was internationally established at 20-30 eggs or three larvae per plant during the first 4-6 weeks after planting (1).

As the cost for control is not accounted for, this decision index does not completely fit the definition of economic
injury level (9).

The use of this injury level for control decisions presu-
mes a cheap, reliable and practicaly usable sampling system.
The egg-trap developed by Freuler and Fischer seems to fulfil
these requirements during their experiments (3).

During 1983 and 1984 the efficiency and selectivity of
the egg-traps were examined. A first analysis of the results
of 1983 was published by Pelerents et al. (8). Here data from
both years were analysed and the number of collected eggs was
compared with the captures of adult flies by water-traps. The
distribution of cabbage root fly eggs in the fields was studied in view of an optimal placing of the traps. Taken into
account the economic injury level, this study focusses on the
captures during the first weeks after transplanting.

2. Experimental design

During 1983 and 1984 the adult population and the oviposition of the first and second generation cabbage root fly
was followed on nine cauliflower fields, scattered in the
Flemish cabbage growing area and covering about 900 km². On
each field, independent of the number of plants, four water
traps (5) were placed at the corners. With the exception of
the fields in Afsnee during 1983 and 1984, ten egg-traps were
placed on the second row perpendicular to the road and from
plant 2 untill plant 11. In Afsnee during 1983 and 1984, 48
egg-traps were placed at regular distances in the field. For
a comparison of the sampling by means of 10 and 48 traps, 10
additional traps were placed in the above described way during
1984. Untill May 15th of both years, the captured females
were dissected and examined for insemination and ovary development. A survey of the size of the fields and the periods of
sampling is given in table I.

3. Results and discussion

3.1. Efficiency of the egg-trap

In 1983 and 1984 respectively 28,860 and 30,711 cabbage
root fly eggs were collected during 2,734/6,074 observations,
giving a mean of 10.5 and 5.1 eggs per trap per observation.
With the egg-trap used earlier (6) the catches average only
2.1 eggs (6,984 observations). This higher efficiency makes
the Freuler trap more useful for the monitoring of the oviposition, although not all eggs layed at the sampled plants were
collected. A root control of the plants which had a Freuler
egg-trap during the whole of the growing season still gave an
average of 5 larvae and pupae per plant.

3.2. Selectivity of the egg-trap

Only a very limited number of eggs from other species
were found in the traps (1983 : 0.6%, 1984: 1.65%). The high
selectivity permits the omission of the identification of the
eggs, so that use by farmers becomes possible.

Table I. Experimental design

Site	No. of cauliflowers	No. of egg-traps	Sampling periods			
			1st gen. 1983	2nd gen. 1983	1st gen. 1984	2nd gen. 1984
Afsnee	2,400	48	11.4-8.9			
	2,400	48 + 10			24.4-28.6	12.7-20.9
Ardooie I	8,500	10	14.4-9.5	27.6-1.8	12.4-21.6	5.7-13.9
Ardooie II	8,500	10	21.4-9.5	27.6-1.8		
	14,000	10			20.4-2.7	5.7-3.9
Asper	18,400	10	21.4-8.9		24.4-18.10	
Bassevelde	120	10		30.6-11.8		
Drongen	50	10	11.4-27.6		24.4-2.7	
Kluizen	50	10	11.4-9.5	30.6-8.9	26.4-2.7	19.7-27.9
Kruishoutem	1,600	10	11.4-9.5			
	2,400	10			24.4-21.6	
	4,900	10				5.7-20.9
Maldegem	50	10	11.4-30.6			
Melle	50	10		4.7-18.8		
Vinderhoute	5,700	10			16.4-21.6	5.7-27.9
Wannegem-Lede	1,000	10	11.4-9.5		26.4-2.7	19.7-18.10

3.3. Determination of the first adults and eggs

Both the water-traps and the egg-traps catch respectively the first adults and eggs. Not a single contradiction was found between these and other observations on the biology such as the emergence of spring collections of diapauzing pupae buried in the soil the previous year (6), the development of the ovaries of the females and the root controls 4-5 weeks after the beginning of the oviposition.

In spring there is a difference in time of about 10 days between the first captures of males and females on the different fields. The first eggs were found on nearly all fields on the same day, 28 April 1983 and 30 April 1984. Three days earlier, 0.1 eggs per trap were observed on 2 fields in both years. In summer cauliflowers males, females and eggs were always collected from the first observation after transplanting.

3.4. Distribution of fly- and egg-population

For the eggs, the relationship between the variance and the mean obeys Taylor's power law $s^2 = a\bar{x}^b$ (10). The regression between the variance and the mean (after log transformation) was calculated for each generation, each year and each field aswell as for the sampling with 48 egg-traps in Afsnee. All regressions were significant (P:0.01) and all regression coefficients were homogenous (4). The regression calculated with all data (Table II) is similar to the one FINCH et al. found (2). In the same way Taylor's power law was calculated for males, females and inseminated females. Besides the captures of 1983-1984, the data gathered in earlier experiments (6) were used.

Table II. Taylor's power law ($\log s^2 = \log a + b \log \bar{x}$) for the cabbage root fly populations

	log a	b	n	t_b	level of significance	r
eggs	0.56	1.50	578	97.01	0.001	0.97
eggs Finch et al.(2)	1.31	1.41				
males	0.47	1.66	175	48.70	0.001	0.93
females	0.13	1.18	159	25.62	0.001	0.90
inseminated females	0.05	1.10	75	14.78	0.001	0.87

According to Taylor (10) b is an index of aggregation characteristic to the species and the stage. The aggregation indices of males and females differ significantly, those for females and inseminated females do not.

For further calculations the data were transformed from x to x^p with p = 1 - 0.5 b (10). So egg data were transformed to $x^{0.25}$, the captured females to $x^{0.40}$. These transformations stabilize the variance well, except for the egg data on a few sites (+ 2%) where a mean population lower than 2 eggs/trap/observation was found.

3.5. The relation females - eggs

The ratio females/eggs differ between the fields e.g. for every female captured in Maldegem 1.9 eggs were found; in Afsnee this ratio was 10 times smaller (18.4-13.5.83). A cause may be the difference in age of the females. In both years, the spermatheca and the ovaries of all females captured untill May 15th were examined. A high correlation coefficient r = 0.92 (significant 0.01) is found between the number of inseminated females and the number of females with terminal oocytes in the ovaries (both after log(x+1)-transformation). When eggs and females captured on the different fields during one observation period are added together and these sums are correlated for the different times, a high correlation coefficient is found (1983: r=0.90; 1984: r=0.91). Replacing the females by the number of inseminated females gives a slight increase of the correlation coefficient (1983: r=0.96; 1984: r=0.92). These correlations disappear when all individual pairs of eggs/females are considered. Replacing the females by the inseminated ones, gives a slight but an insufficient increase. The correlation between eggs and females with terminal oocytes gives similar results.

When all females and eggs captured during 1983 and 1984 are transformed respectively to $x^{0.4}$ and $x^{0.25}$, the correlation between both stays low and differs for both years. The correlation calculated only with the captures of the first generation during the first five weeks after transplanting is higher (r=0.68) and the regression coefficients of 1983 and 1984 are homogenous. One regression with confidence and prediction-limits (7) is calculated with the data of both years and is

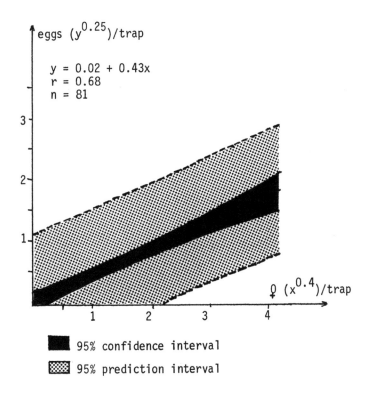

eggs $(y^{0.25})$/trap

y = 0.02 + 0.43x
r = 0.68
n = 81

♀ $(x^{0.4})$/trap

■ 95% confidence interval

▒ 95% prediction interval

Figure 1. Relationship females/eggs - first
generation - first five weeks after
transplanting

represented in figure 1. An accurate prediction of the eggs on
the bases of female captures is not possible as long as the
relationship females/eggs is not clear.

3.6. Place and number of egg-traps

During four field experiments in Afsnee, 48 egg-traps were
placed in fields of about 3,000 cauliflowers. An analysis of
variance and Duncan's multiple range test (4) for the eggs,
transformed to $x^{0.25}$, shows beside a significant influence of
time, differences between the catches in the different traps.
In comparison with the mean, the traps less than 3 m x 10 m
away from any corner, always catch an equal or significant
higher number of eggs (figure 2.). In agreement with earlier
conclusions the highest number of eggs were caught adjacent
to a road (8).
A comparison of the recaptures in the ten egg-traps at a
corner and the 48 at regular distances in the field is given
in figure 3. The correlation coefficient between both (after
transformation $x^{0.25}$) is 0.86.

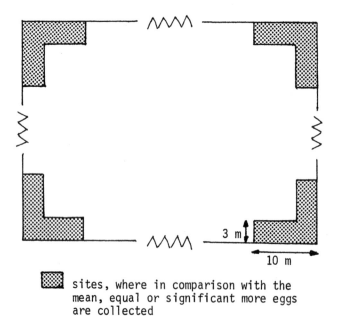

sites, where in comparison with the
mean, equal or significant more eggs
are collected

Figure 2. Optimal positioning of egg-traps

4. Conclusions

    The high efficiency and selectivity make the Freuler egg-
traps very suitable for research aswell as for monitoring purposes. As the period when the first eggs are found does not
differ too much one year to another, the sampling system used
should be rather precise so as to be useful. Although both
water- and eggs-traps catch respectively the first adults and
the first eggs, the prediction of the start of the oviposition
on the basis of female captures is too inaccurate to be of any
use.
    With ten egg-traps placed at a corner of the field and if
possible near a road, the first eggs are detected. The risks
of missing the eggs is very small. Taking into account the
economic injury level an additional prevention is built in
when the cabbage root fly is controled the moment the first
eggs are found. At the same time room is left for a delay of a
few days between the detection of the eggs and application of
control measures.
    This monitoring system will be especially useful in very
early and early cauliflowers, which often are treated unneces-
sarly. Even when control is needed, smaller doses of insecticides can be applied at the optimal moment so that the side-
effects on beneficial organisms are minimised.

72

First generation - Afsnee - 1984

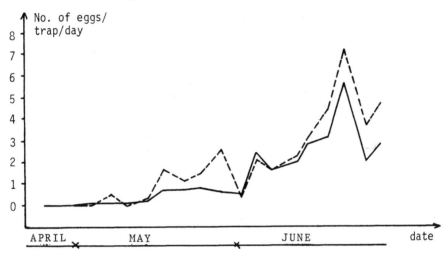

Second generation - Afsnee - 1984

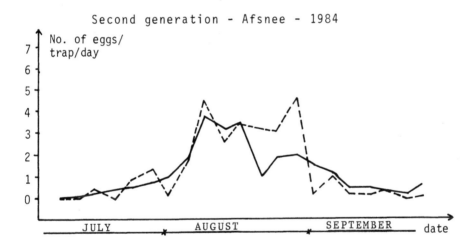

Figure 3. Influence of number and place of the egg-traps
on the average number of eggs sampled (--- 10
traps at one corner; ——— 48 traps in the field).

Acknowledgements

This research was supported by the Institute for the
Encouragement of Scientific Research in Industry and Agricul-
ture (IWONL-IRSIA). The authors thank G. Buysse, P. Pelerents,
I. De Vloedt and R. Van Caenegem for technical assistance.

73

References

1. COAKER, T.H. (1979). Integrated control in brassica crops. IOBC-WPRS Bulletin II/4.
2. FINCH, S., SKINNER, G. and FREEMAN, G.H. (1975). The distribution and analysis of cabbage root fly egg populations. Ann. appl. Biol., 79, 1-18.
3. FREULER, J. and FISCHER, S. (1983). Le piège à oeufs, nouveau moyen de prévision d'attaque pour la mouche du chou, *Delia radicum (brassicae)* L.. Revue Suisse Vitic. Arboric. Hortic. 15(2): 107-110.
4. GOMEZ, K.A. and GOMEZ, A.A. (1976). Statistical procedures for agricultural research. IRRI, Los Banos, Philippines, pp. 282.
5. HAWKES, C. (1969). The behavior and ecology of the adult cabbage root fly *Erioischia brassicae* (Bouché). Ph.D. Thesis, University of Bermingham.
6. HERTVELDT, L., VAN KEYMEULEN, M. and PELERENTS, C. (1980). Development of the sterile insect release method against the cabbage root fly, *Delia brassicae* B. in North Belgium. IOBC-WPRS-Bulletin III/1:63-87.
7. MENDENHALL, W. (1979). Introduction to Probability and Statistics. Wadsworth Publishing Company, California, pp. 594.
8. PELERENTS, C. and VAN KEYMEULEN, M. (1984). Methods for the Monitoring and Integrated Control of Cabbage Root Fly *Delia radicum (brassicae)*(L.). EUR 8689 - Final report of the CEC 'Integrated and Biological control' Programme 1979-1983, 273-285.
9. POSTON, F.L., PEDIGO, L.P. and WELCH, S.M. (1983). Economic Injury Levels: Reality and Practicality. Bull. Ent. Soc. Am. 29 (1), 49-53.
10. TAYLOR, L.R. (1961). Aggregation, variance and the mean. Nature, London 189, 732-735.

Effect of chemical treatment on *Delia radicum* and its parasitoids *Aleochara bilineata* and *Trybliographa rapae*

B.Bromand
Institute of Pesticides, National Research Centre for Plant Protection, Lyngby, Denmark

Summary

Field trials in 1985 clearly showed that seedbed treatment of cauliflower with chlorfenvinphos 1 g a. i. per m^2 had the best effect on Delia radicum larvae followed by carbofuran granules 0,05 g a.i. and carbosulfan 0,075 g a.i. per m row at transplanting.

When applied sowing soil treatment with chlorfenvinphos 4 kg a. i. per ha, carbofuran granules 0,05 g a. i. and carbosulfan 0,075 g a. i. per m row had the best effect. Seed treatment and spraying gave unsatisfactory results.

The number of D. radicum pupae were greatly reduced by chlorfenvinphos (65%), carbofuran (51%) and carbosulfan (38%) when chlorfenvinphos was used as soil treatment.

The number of pupae parasitized by Aleochara bilineata and Trybliographa rapae dropped from 78% in untreated to 57% with chlorfenvinphos and 66% with carbosulfan.

1. Introduction

The cabbage rootfly is a very serious pest on cabbage in Denmark. In spring the cabbage is transplanted but later during the summer the plants in some fields are drilled.

The first generation of the cabbage rootfly starts around the middle of May and last for about 1 month. The second generation starts in the beginning of July and last for about 6 weeks. According to the number of cabbage rootfly eggs laid in egg traps the second generation is 10 fold more numerous than first generation.

This is partly believed to be dependent on the increasing area of oilseed rape. In oilseed rape the first generation multiply enormously but when second generation hatches the rape plants are not atractive for egglaying anylonger and the flies seek out cabbage fields, where they are becoming an increasing problem.

No satisfactory chemical treatment exist, which leads to an excessive use of chemicals. It is important to find out the best way of using the existing chemicals at the optimal time in order to obtain the best effect and at the same time the least damage to beneficial arthropods.

In 1985 4 field trials were carried out with soil treatment, soil incorporated granules, spraying and seed-bed treatment. The results were evaluated on the number of useful cabbage produced as well as on propagation of cabbage rootfly pupae and parasitization of the pupae with A. bilinata and T. rapae. Only the results from 2 trials with heavy attacks are included.

2. Materials and methods

The trials were carried out in cauliflower after different plans the first of which can be seen in table I.

Table I. Plan over field trials with transplanted cabbage

Treatment	Method	Dosage
1. Untreated		
2. Curaterr (carbofuran 5%)	granules	1 g per m row
3. Sumicidin 10 FW (fenvalerat 10%)	spraying	0,6 l per ha
4. Marshall 5 G (carbosulfan 5%)	granules	1,5 g per m row
5. Orthene 75 SP (acephat 75%)	spraying	1,5 kg pr ha
6. Shell Birlane 24 EC (chlorfenvinphos 24,7%)	seed-bed treatment	4 g pr m^2
7. Trigard 75 WP (cyromazin 75%)	spraying	0,2 kg per ha

The cabbage was transplanted on May 9th and spraying took place June 2nd. Seed-bed treatment took place just before transplanting of the cabbage. After the treatment the plants were sprinkled with water. Twice in June the number of plants killed by the cabbage rootfly larvae were counted and later the number of heads larger than 12 cm in diameter were counted. This plan was carried out at an appropriate time according to the first generation of the cabbage rootfly.

In two trials the cabbage was drilled June 6th, allowing the plants to get a reasonable size at the time of egglaying of second generation and at the same time to give the seed treatments a reasonable change to have an effect. The trials were carried out according to the plan in table II.

Table_II._Plan_over_field_trials_with_sown_cabbage

Treatment	Method	Dosage
1. Untreated		
2. Oftanol bejdse (isofenphos 40%)	seed treatment	40 g per kg
3. Promet Twin 47,5 SD		
(furathiocarb 40%)	seed treatment	40 g per kg
4. Promet Thiram 47 SD		
(furathiocarb 40%)	seed treatment	40 g per kg
5. Rapcol 49,5 SD (furathiocarb 40%)	seed treatment	40 g per kg
6. Trigard 75 WP (cyromazin 75%)	seed treatment	25 g per kg
7. Trigard 75 WP (cyromazin 75%)	seed treatment	25 g per kg
Trigard 75 WP (cyromazin 75%)	spraying	0,2 kg per ha
8. Trigard 75 WP (cyromazin 75%)	spraying	0,2 kg per ha
9. Shell Birlane 24 EC		
(chlorfenvinphos 24,7%)	soil treatment	16 kg per ha
10. Furadan 5 G (carbofuran 5%)	granules	1 g per m row
11. Trigard 75 WP (cyromazin 75%)	soil treatment	1,3 kg per ha
12. Marshall 10 G (carbosulfan 10%)	granules	0,75 g per m row

Several of the products contains fungicides as well as insecticides. Fungicides are not mentioned in table II, as they have no
effect on the cabbage rootfly.
The sprayed plots were treated July 22nd and August 7th. The
number of plants killed by the cabbage rootfly were noted and at
cutting the number of with useable cabbages were caunted (heads
larger than 12 cm in diameter. September 30th soil samples were
taken around 5 plants per plot and the number of cabbage pupae
were registered. The pupae were then investigated for parasitization of A. bilineata and T. rapae.
All the trials were carried out with 4 replicates and at spraying
1000 l of water per ha was applied.

3. Results

The results from one trial with transplanting can be seen in
table III.
The attack by the cabbage rootfly started around May 23rd about
8-10 days later than usual due to the cold spring. In the beginning of June Curaterr, Marshall 5 G and Shell Birlane 24 EC
showed good effect on the number of plants but later the effect

Table III. Per cent dead plants and useable heads

| Treatment | Per cent dead plants | | Per cent |
	June 4th	June 19th	useable heads
1. Untreated	29	71	18
2. Curaterr	3	5	82
3. Sumicidin 10 FW	25	65	23
4. Marshall 5 G	3	20	58
5. Orthene 75 SP	26	60	23
6. Shell Birlane 24 EC	2	6	90
7. Trigard 75 WP	24	69	20

of Marshall 5 G was weaker. The same was the case with the number of usable heads. Spraying was ineffective.

The same results were achieved in the trials where the cabbage was sown, but with less pronounced effect. See table IV.

Table_IV.__Per_cent_dead_plant_and_useable_plants

Treatment	Per_cent_dead_plants at cutting	Per_cent usable heads
1. Untreated	32	24
2. Oftanol bejdse	26	50
3. Promet Twin 47,5 SD	24	32
4. Promet Thiram 47 SD	36	28
5. Rapcol 49,5 SD	36	32
6. Trigard 75 WP	36	28
7. Trigard 75 WP +	48	28
Trigard 75 WP	32	20
8. Trigard 75 WP		
9. Shell Birlane 24 EC	4	64
10. Furadan 5 G	20	68
11. Trigard 75 WP	28	32
12. Marshall 10 G	12	56

Seed treatments in general had poor effect. The only positive effect was obtained with Oftanol bejdse with 50 per cent useable heads.

The effect of the chemicals on the number of pupae in the soil is shown in table V.

The number of pupae in the soil was greatly reduced by Shell Birlane 24 EC and Furadan 5 G and less by Marshall 10 G. Soil treatment and spraying with Trigard 75 WP and seed treatment with Rapcol 49,5 SD and Trigard 75 WP showed less effect.

Table V. Number of D. radicum pupae and percent pupae parasitized by A. bilineata and T. rapae.

Treatment	Number of pupae		Per cent parasitized		
	Total	Relative	Total	A.bilineata	T.rapae
1. Untreated	235	100	78	23	55
2. Oftanol bejdse	234	100	82	18	64
3. Promet Twin 47,5 SD	248	105	90	26	64
4. Promet Thiram 47 SD	274	117	88	26	62
5. Rapcol 49,5 SD	185	79	81	25	56
6. Trigard 75 WP	205	87	85	22	63
7. Trigard 75 WP + Trigard 75 WP	183	78	82	21	61
8. Trigard 75 WP	254	108	84	27	57
9. Shell Birlane 24 EC	83	35	57	9	48
10. Furadan 5 G	115	49	78	14	64
11. Trigard 75 WP	213	91	89	21	68
12. Marshall 10 G	145	62	66	16	50

The percentage of parasitigation by A. bilineata and T. rapae was reduced by Shell birlane 24 EC soil treatment and Furadan 5 G and Marshall 10 G granules despite of the lower number of pupae in these plots. That means that these methods and chemicals have an even more adverse effect on these parasitoids.

4. Discussion and conclusion

Although the results only came from a limited number of trials they show some promising results, but more trials are needed. Especially the seed-bed treatment should be further investigated.

How long time before transplanting should the treatment of the plants take place and what is the least effective dosage ? Can seed-bed treatment be used with cabbage plants in peat blocks ? What is the safety aspects for people handling the plants during transplanting ? Trials with chinese cabbage have shown that no chemicals were left in the plants at harvest and this presumably goes for cauliflowers too.

First generation of the cabbage rootfly last for about one month. Granules at transplanting usually gives a fairly good control of this generation. In practice however it has been demonstrated that the effect of carbofuran granules last for no more that 4 weeks and this is not sufficiant for control of second generation.

In this case good results have been obtained by watering the plants with 0,1 l per plant of a solution of diazinon.

Root damage indexes were calculated in the trials too, but the results are not brought here because there were only slight differences between treatments, obviously because the indexes were calculated on surviving plants only.

The different control measures and treatments have demonstrated different effect on parasitization of A. bilineata and T. rapae.

However it is difficult to get the exact information about this. The pupae came from surviving plants only and you find a different number of pupae in different treatments. This means a greater percentage of pupae should become paratized when fewer pupae are found in order to have the same population of parasites and of predators too.

From these trials can be concluded that seed-bed treatment with 1 g a.i. of chlorfenvinphos per m^2 seams to be the most effective control measure for transplanted cabbage against the cabbage rootfly.

Development of various formulations for controlled release of naphthalene as a repellent against oviposition of the cabbage root fly, *Delia radicum (brassicae)*

H.den Ouden
Research Institute for Plant Protection, Wageningen, Netherlands

Summary

Clay loaded carboxymethylcellulose and plaster of Paris granules have been manufactured as substitutes of initially tested synthetic polymers as controlled release matrices for the repellent naphthalene. The formulations loaded with different percentages of active ingredient have been used in experiments in fields with cauliflower, attacked by the cabbage root fly, Delia radicum (brassicae). The results were compared with the effectivity of a conventional insecticide treatment.

1. INTRODUCTION

It has been shown that the possibilities for application of repellents against D. radicum are limited by the necessity to use large quantities of a controlled release formulation with a low quantity of active ingredient (den Ouden et al., 1984; den Ouden, 1984). The period of activity must be at least six weeks. The problems are caused mainly by the high volatility of the available repellents which prevent the use of very small particles. As a consequence, quantities of approximately 500 kg of formulated product per ha should be spread around the stems in the cabbage field. The costs of the degradable polymers available for this purpose have risen considerably during the last years and therefore experiments have been carried out with granules of clay matrices for the cheapest synthetic repellent available, i.e. naphthalene. These formulations can be used only against the cabbage root fly when it is attacking the root system. They are useless for control of oviposition and maggots in the heads of cabbage and buttons of Brussels sprouts.

2. MATERIALS AND METHODS

For the production of granules with a sufficiently long retention time of the active ingredient, two parameters are important: stability and size. When materials such as clay powder and fodder wheat flour are used for the granules, these should be protected against degradation, e.g. caused by rainfall. As a stabiliser, carboxymethylcellulose (CMC) has been chosen being a relatively cheap and degradable polymer which can be mixed easily with clay or flour. For additional stability a hydrophilic emulsion of the hydrophobic polyvinylacetate (PVAc) has been applied.

For an experiment in 1984, 10 kg clay powder or flour was mixed initially with 1600 g and/or 800 g of ground naphthalene flakes. This mixture was added to 7 l of gel containing 7.5% of CMC and 3.8% of PVAc emulsion and spread as a 1 cm thick layer in a tray for a drying procedure of circa 10 days at 21°C. The concrete-like plate was then reduced easily into fragments of circa 0.75 cm.

More simple was the manufacturing of granules from plaster of Paris in 1985. Powder of this substance was mixed with 7, respectively 15%, of naphthalene powder. Water was added to a quantity of circa 50% of this mixture by spraying it in very fine droplets onto the mass which was turning around in a concrete mixer until granules of an average diameter of 4 mm had been formed. These were spread on a tray and dried in circa 12 hours. Irregularly shaped granules might have been manufactured by crushing dry plates of plaster of Paris, cast onto a plastic sheet 12 hours before.

As it has been shown that naphthalene has no effect at a distance of more than a few cm (den Ouden et al., 1984), treated and untreated plots can be placed close to each other. The field experiments were laid out on sandy soil in rectangular fields where three randomised blocks of a certain number of plots functioned as replicates. Per plot 7 x 7 cauliflowers var. Alpha were planted in the first 10 days of May (147 plants per treatment). Circa 10 days later the plants were provided with varying amounts of repellent granules around the stem base. As a standard treatment trichloronate was added (100 ml 0.1% Phytosol per plant) and pure naphthalene (twice 1, 3 or 6 g per plant on May 5 and June 1) in 1984 only.

Root damage indices (RDI) were determined according to a standard of the IOBC/WPRS Brassica Working Group (Thompson, 1980). The indices varied from 1 (no damage by maggots in the root) to 4 (seriously damaged root). Significance of differences was determined by the χ^2-test. Plants with RDI 3 and 4 were considered to yield an inferior product.

In the second half of the season 1985 an experiment where plaster of Paris granules were applied was hardly or not infected by the cabbage root fly. In this experiment the curds were harvested and the yields compared with those from the trichloronate-treated plots.

3. RESULTS AND DISCUSSION

Table 1 shows the results of the experiment in 1984. There are many significant differences but here the not significant differences with the generally almost 100% effective trichloronate application are more interesting. The first three formulations do not differ very much in this respect although clay with 15% and flour with 7.5% naphthalene seem to be somewhat better. A remarkable phenomenon is the absence of a dose effect. This may have been caused by the relatively small differences at the dosages which have been chosen but also by the fact that D. radicum lays its eggs preferably very near to the stem. This may cause a good repellent action with even small dosage.

The accurate assessment with the RDI-system showed clay granules with a load of 7.5% a.i. to be inferior to the 15% loaded granules and the flour with 7.5% naphthalene. This applies for the 20 and 30 g dosage.

Strikingly low is the effect of the pure naphthalene which has been applied twice around the stems of the cauliflowers. If it had been applied not as flakes but as coarse particles, e.g. Caulin rings (de Wilde, 1947), the effect would have been somewhat better by a slower evaporation.

Generally the variance appeared to be considerable and it was because of the individual observation of each plant and because of the large numbers per treatment that reliable figures could be obtained.

The same applies for the 1985 experiment. Table 2 shows that plaster of Paris granules are just as effective as the clay and flour formulations.

TABLE 1. The Repellent Effect of Various Formulations of Naphthalene on Stand and Root Damage of Cauliflowers Assessed Individually
Stand scores ranging from 0-2. Root damage indices ranging from 1-4.
Figures with - and · do not differ from the treatment with Phytosol 100 ml at P > 0.05 and P > 0.02 respectively.

Formulations	Clay + 7.5% a.i.			Clay + 15% a.i.			Flour + 7.5% a.i.			Pure a.i., 2x			Untreated -			Phytosol 0.1%		
Dosage in g	10	20	30	10	20	30	10	20	30	1	3	6	-	-	-	25	50	100
Percentage plants with stand 2 at July 5	87·	90⁻	82	91⁻	90⁻	86·	93⁻	78	87⁻	58	61	60	45	64	68	94	75	94
Percentage plants with RDI 1+2 at July 9	87	78	68	88	93⁻	90·	94⁻	88	90·	71	59	72	58	77	72	99	90	97
Average block total stand																		
May 30	98	99	95	98	98	98	98	98	98	97	97	97	98	96	97	98	95	98
June 28	89	95	88	92	93	86	93	85	89	73	74	71	63	78	77	93	88	92
July 5	90	92	86	93	92	89	94	85	89	70	72	70	60	76	74	95	83	94

TABLE 2. The Effect of Various Dosages of Naphthalene Formulated in Plaster of Paris on the Infestation Rate and RDI in 3 Plots of Cauliflower Plants

Formulations with % active ingredients	Plaster of Paris + 7% a.i.	Plaster of Paris + 15% a.i.	Plaster of Paris + 7% a.i.	Plaster of Paris + 15% a.i.	Untreated - -	Trichloronate Phytosol 0.1%
Dosage in g	10	10	20	20	-	100
Percentage of plants seriously attacked						
at May 28	0	0	0	0	9	0
at June 4	1	1	5	1	32	0
at June 11	3	1	7	2	37	0
Percentage of plants with RDI 1+2 at June 21	77	91	81	90	22	98

There is a significant RDI difference here between the 7% and the 15% load; for the 10 g dosage at P < 0.01, for the 20 g dose at P < 0.05. The RDI differences between the naphthalene formulations and the trichloronate treatment are significant at P < 0.01. The percentages of seriously attacked plants after treatment with 10 and 20 g of granules with 15% only did not differ significantly (P > 0.05) from trichloronate treatment.

In addition to the 1984 field experiment with the clay and flour granules a GLC test was done at the Department of Organic Chemistry of the Agricultural University, Wageningen. It was shown that the drying procedure of the granules caused a loss of circa 23% of the naphthalene. Granules kept in a field depot appeared to have released another 50% from May 10 to June 6 and the load had decreased to less than 3% at June 29.

The good results of this way of control of the cabbage root fly have been obtained not only by the repellent effect of the naphthalene formulations but also by the increasing tolerance of the plants and the decrease of the population density of the fly. It is obvious that the accurate judging by the RDI method generally gives a worse impression than when the general stand of the crop is observed. This does not apply for the insecticide treatment, which evidently has a more prolonged effect.

It should be emphasised that the cheap formulations described demand for an adapted machine for eventual application in the commercial cultures. In private gardens hand application will provide an easy means of control of the cabbage root fly. The granules should be stored cool because of sublimation of the active ingredient.

Finally the result of the comparison between trichloronate treated and naphthalene-plaster of Paris treated crop should be mentioned. In a test with a total of 512 curds, there was no significant difference ($P \gg 0.1$) in the ratios between the percentages of curds belonging to grades I + II (different somewhat in shape only) and the curds of grade III being of an inferior quality and size.

ACKNOWLEDGEMENTS

Thanks are due to Dr. G.M. Sanders and Mr. E.J.C. van der Klift of the Department of Organic Chemistry of the Agricultural University, for carrying out the naphthalene analyses in clay granules, to Mr. J.L. Kalthoff of Enka bv, Arnhem, for providing a CMC supply and to Messrs. J. Freriks, J. van der Beek and S. Folkertsma for assistance in the field experiments.

REFERENCES

(1) OUDEN, H. DEN, 1984. Untersuchungen über Repellents im Einsatz gegen die kleine Kohlfliege. Mitt. a.d. Biol. Bundesanst. für Land und Forstwirtschaft 218: 51-59.
(2) OUDEN, H. DEN, THEUNISSEN, J. and HESLINGA, A., 1984. Protection of cabbage against oviposition of cabbage root fly Delia brassicae L. by controlled release of naphthalene. Z. ang. Ent. 97: 341-346.
(3) THOMPSON, A.R., 1980. The assessment of damage caused to cruciferous crops by larvae of Hylemya brassicae in soil. WPRS Bulletin III/1: 19-26.
(4) WILDE, J. DE, 1947. Onderzoekingen betreffende de koolvlieg en zijn bestrijding. Staatsdrukkerij 's Gravenhage 1947.

The control of cabbage root fly, *Delia brassicae* (Wied.), in transplanted brassicas by insecticide application to the seedbed – A report of collaborative work

D.J.Mowat

Agricultural Zoology Research Division, Department of Agriculture for Northern Ireland, and Department of Agricultural Zoology, Queen's University of Belfast, Northern Ireland

Summary

A research programme was initiated to confirm the effectiveness of, and extend familiarity with, a method of cabbage root fly control which involves minimal use of insecticide. Several workers throughout Europe co-operated. The method requires the application of chlorfenvinphos or chlorpyrifos to a brassica seedbed at 10 or 20 kg ha^{-1} and is relevant only to brassicas raised in a seedbed and transplanted. Where cabbage root fly attack was sufficient to affect growth, yield increases of up to 303% were recorded. Seedbed treatments compared favourably with standard treatments, where these were included, and in one case significantly outyielded a standard treatment. Reduction of root damage was generally only slightly less in seedbed treatments than in standard treatments.

Seedbed treatment, previously used mainly in summer cauliflowers, compared favourably with standard treatments in protecting brussels sprout crops, which have a longer growing season and, therefore, test the persistence of effectiveness more fully.

Seedbed treatment with chlorpyrifos did not adversely affect parasitism of cabbage root fly larvae and pupae, but parasite numbers were reduced by corresponding chlorfenvinphos treatments and by field treatments.

Two experiments, in which plants were raised in peat blocks, confirmed a previous observation that insecticide application rate can be reduced by following the principles of the seedbed method, notably by earlier post-emergence application.

1.1 Introduction

The protection of transplanted brassica plants from cabbage root fly attack in the field was shown to be possible by treating only the seedbed with insecticide (2). The transfer of seedbed soil on plant roots was not essential, and the main requirements were an adequate time interval between insecticide application and transplanting to allow impregnation of the roots, and an insecticide of at least moderate persistence. To ensure effectiveness relatively high application rates were used initially (3) but continuing study showed that consistently satisfactory results were obtained at lower rates (6). As the introduction of insecticide to the field was negligible an adverse effect on the predators of cabbage root fly eggs was unlikely and it was confirmed that seedbed treatment and predatory beetles gave useful complementary effects (7). At the Working Group meeting in Dublin, 1982 it was agreed to initiate a collaborative project to extend knowledge of this control method.

1.2 Methods

Suggested treatments were chlorfenvinphos or chlorpyrifos, in liquid or granular formulation, applied to the seedbed at 10 or 20 kg ha^{-1}. Application times were before sowing (which was not recommended where plants were likely to remain in the seedbed for more than 2 months) or post-emergence. For post-emergence treatments an interval of 3-4 weeks between treatment and transplanting was recommended for drench applications and 5 weeks for granule applications. The concentration of post-emergence drenches did not exceed 0.1% chlorfenvinphos or 0.2% chlorpyrifos.

Collaborators were asked to include some or all of these treatments, with a no-insecticide treatment and, where possible, at least one standard field treatment for comparison. All used randomised block designs.

Suggested assessment methods were root damage during the growing season and at harvest, and yield. The resources of collaborators varied, so that the number of treatments and assessments also varied within the prescribed framework. Eight workers collaborated but only five experiments from three centres were quotable, owing to inadequate cabbage root fly attack or inappropriate treatments in some cases.

1.3 Results

Cabbage

At Cambridge, England the cultivar Primo received 10 kg chlorfenvinphos or chlorpyrifos ha^{-1} (e.c. formulation only), applied post-emergence (29 May, 1983), 31 days before transplanting. At harvest (5 August) 37 days after transplanting, the treatments had reduced root damage by 69% on average ($P < 0.05$) and increased yield by 138% (chlorfenvinphos) and 115% (chlorpyrifos) (Table I). The crop was exposed to first and second generation attack. In an adjacent experiment the first generation attack was low but the second was high and it can be assumed that part of the effect was due to control of the second generation. Yield was low overall, due to poor soil and lack of irrigation except at planting, but the quality of cabbages was good.

Cauliflower

At Wellesbourne, England the cultivar Perfection received all 16 of the recommended treatments as well as four standard field treatments for comparison. Post-emergence treatments (27 April, 1983) were applied 14 days before transplanting (11 May). All seedbed soil was washed from the roots before planting. Root damage assessment on 21 June (Table I) showed that all e.c. treatments had reduced damage ($P < 0.01$), as had all granule treatments at the higher rate. Granules applied at the lower rate before sowing also reduced damage significantly. Spot applications in the field were generally a little more effective but the best seedbed treatments compared favourably and chlorpyrifos post-emergence drenches were the most effective treatments overall.

At Loughgall, Northern Ireland the cultivar Dok received the four recommended pre-sowing treatments at 20 kg ha^{-1} and the post-emergence drench treatments (17 May, 1983) at 10 kg ha^{-1}. Transplanting was on 29 June and the attack was severe. The percentage of plants which were marketable, at harvest from 7 September to 11 October, was increased ($P < 0.01$) by all treatments (Table I). The seedbed treatments outyielded the control (in which many plants were killed by cabbage

Table I   Root damage by cabbage root fly and yield of cabbage and cauliflower plants following insecticide application in the seedbed or in the field.

	Cambridge Cabbage (Primo) R.D.I.	Cambridge Cabbage (Primo) Yield (kg/20 plants)	Wellesbourne Cauliflower (Perfection) R.D.I.	Loughgall Cauliflower (Dok) % marketable
Untreated	67(±6.9)	5.5(±1.2)	56	18.0
Seedbed treatment (kg a.i./ha)				
chlorfenvinphos				
pre-sowing e.c. (10)	-	-	38	-
(20)	-	-	34	65.0
granules (10)	-	-	42	-
(20)	-	-	35	82.1
post-emergence				
e.c. (10)	15(±0.4)	13.1(±2.2)	36	72.7
(20)	-	-	38	-
granules (10)	-	-	49	-
(20)	-	-	31	-
chlorpyrifos				
pre-sowing e.c. (10)	-	-	39	-
(20)	-	-	33	70.0
granules (10)	-	-	36	-
(20)	-	-	31	79.6
post-emergence				
e.c. (10)	26(±8.4)	11.8(±0.9)	23	65.7
(20)	-	-	19	-
granules (10)	-	-	44	-
(20)	-	-	36	-
Field treatments (mg a.i./plant)				
chlorfenvinphos				
e.c. (31)	-	-	31	-
granules (17)	-	-	23	-
chlorpyrifos				
e.c. (34)	-	-	30	-
granules (17)	-	-	26	-
diazinon				
granules (45)†	-	-	-	53.6
S.E.			4.5	8.54

† Plants also received diazinon granules at 5 kg a.i./ha in the seedbed before sowing

Table II Root damage by cabbage root fly and yield of brussels sprout plants (cv. Hal) following insecticide application in the seedbed or in the field at Loughgall, Northern Ireland

	1983	1984			
	Yield (kg/plot)	R.D.I (6/9/84)	Pupae (1/2/85) per 10 plants	% parasitised	Yield (kg/plot)
Untreated	8.5	37	54	29	18.7
Seedbed treatments (kg a.i./ha)					
chlorfenvinphos					
pre-sowing e.c. (20)	9.8	26	16	6	23.1
granules (20)	10.8	25	19	15	22.9
post-emergence e.c. (10)	13.4	25	36	15	21.5
chlorpyrifos					
pre-sowing e.c. (20)	11.0	23	14	40	19.7
granules (20)	12.9	23	42	27	25.7
post-emergence e.c. (10)	13.6	22	36	36	17.9
Field treatments (mg a.i./plant)†					
diazinon granules (45)	-	27	27	17	21.4
fonofos (40) + disulfoton (60) granules	13.4	6	2	-	21.5
S.E.	1.51	2.8	9.0	6.6	2.47

† Plants also received diazinon granules at 5 kg a.i./ha in the seedbed, before sowing

root fly) by 303% on average and also outyielded a standard diazinon spot treatment by 35% on average - significantly so (P<0.05) in the case of the two pre-sowing granule treatments.

Brussels sprouts

In two experiments at Loughgall, Northern Ireland the cultivar Hal received the four recommended pre-sowing treatments at 20 kg ha^{-1} and the post-emergence drench treatments (17 May, 1983 and 30 May, 1984) at 10 kg ha^{-1}. Transplanting was on 29 June (1983) and 22 June (1984). At harvest, in the first experiment, from 18 to 25 November, 1983 yield increases of 15% to 60% (Table II) were associated with the insecticide treatments. Post-emergence seedbed drenches and the standard field treatment differed (P<0.05) from the untreated. Insecticide treatments did not differ from each other.

In the second experiment (1984) root damage assessment on 6 September (76 days after transplanting) showed reductions (P<0.001) by all treatments (Table II). The standard diazinon/fonofos/disulfoton treatment (Table II) was more effective than any other treatment (P<0.001). All seedbed treatments compared favourably with the other standard treatment (diazinon). Yield, from 10 to 17 December, was not significantly affected by treatment but was 15% higher in treated plots overall than in untreated plots. Reductions (P<0.05) in the number of pupae recovered from the soil by flotation on 1 February 1985 were associated with the standard treatments and some seedbed treatments. The proportion of pupae which were parasitised was reduced by the standard diazinon treatment and by the chlorfenvinphos seedbed treatments but not by the chlorpyrifos seedbed treatments. The chlorfenvinphos treatments collectively differed (P<0.001) from the corresponding chlorpyrifos treatments in this respect. Parasites which emerged from the pupae were almost exclusively *Trybliographa rapae* (Westw.)

1.4 Discussion

In addition to extending successful experience of this method of cabbage root fly control to other centres the results of these experiments added to existing information in several respects.

Previously (3, 4, 5, 6) there have been indications, based on yield or plant survival or quality, that cabbage root fly control by seedbed treatment may be superior to some standard methods of control, as well as being simpler and more economical. In these cases, however, either the plants were not retained until harvest or the root fly attack was insufficient to produce conclusive differences between treatments. In two of the experiments quoted here - cabbages at Cambridge and cauliflowers at Loughgall (Table I) - yield was greatly reduced in untreated plots, very large increases were recorded in seedbed-treated plants and (at Loughgall) seedbed-treated plants outyielded plants which received a standard diazinon treatment in the seedbed and in the field, in some cases significantly. In all experiments in which yield was recorded and in which a standard treatment was included for comparison, seedbed treatments were at least comparable with standard treatments. In the brussels sprout experiments (Table II) there were no significant differences between yields of insecticide treatments, although in both cases insecticide treatments in general outyielded untreated plots. Over the two experiments yields from seedbed-treated and field-treated plots were almost indistinguishable.

The extension of seedbed treatment treatment to the brussels sprout

Table III Root damage by cabbage root fly and yield of cauliflowers grown in peat blocks at Saillon, Switzerland

	1983 cv. Nevada		1984 cv. Celesta	
	R.D.I.	Yield (kg/plot)	R.D.I.	Yield (kg/plot)
Untreated	53.5	9.8	54.8	11.0
Peat block treatments (kg a.i./ha)†				
chlorfenvinphos e.c. (10)	39.9	20.5	–	–
chlorfenvinphos granules (10)	41.7	23.4	–	–
(20)	–	–	39.8	15.2
(40)	–	–	42.9	14.8
chlorpyrifos e.c. (10)	39.5	18.7	–	–
chlorpyrifos granules (10)	44.0	17.5	–	–
(40)	–	–	22.7	14.6
fonofos granules (10)	38.0	20.2	–	–
(40)	–	–	33.5	14.9
carbofuran granules (40)	–	–	22.7	15.6
Field treatment (mg a.i./plant)				
chlorfenvinphos granules (25)	13.9	23.2	22.5	16.7
S.E.	3.94	1.76	7.13	2.38

† Application rates are given per unit area for comparison with seedbed treatments. 10 kg/ha (or 1 g/m²) is equivalent to 1.6 mg/4 cm x 4 cm block.

crop is also an advance on previous published work. As the brussels sprout crop has a longer growing season in the field it provides a more rigorous test of the persistence of control by this method.

Root damage assessment 37 days after planting at Cambridge (Table I) showed reductions associated with seedbed treatment roughly in line with previous published evidence. At Wellesbourne, 41 days after planting (Table I), the reductions in root damage were less than expected. In this case all seedbed soil was removed from the roots and the interval between post-emergence applications and transplanting was shorter than recommended, especially for granule treatments, reducing the opportunity for impregnation of the roots with insecticide. Post-emergence granule treatments reduced damage by 29%, pre-sowing treatments at the lower rate by 31%, all other seedbed treatments by 44% and field treatments by 51%. At Loughgall in 1984 (Table II) root damage reductions 76 days after transplanting averaged 35% by six seedbed treatments and 27% by the standard diazinon treatment. The much greater effectiveness of the diazinon/fonofos/disulfoton treatment was clearly unnecessary as it was not reflected in yield improvement.

The effectiveness of seedbed treatment results mainly from very high levels of damage reduction during the early stages of post-planting growth, as demonstrated by Mowat and Martin (6) and largely confirmed in the Cambridge experiment. Elimination of root damage later in the season is not expected. The recovery of pupae in autumn has shown residual effectiveness throughout the season, following seedbed treatment at rates similar to (2) or greater than (3, 5) those used above. At Loughgall in 1984 (Table II) the number of pupae present in February was reduced by 50% by the seedbed treatments on average, demonstrating continued effectiveness of treatment throughout the period of cabbage root fly activity. Chlorpyrifos treatment did not reduce the level of parasitism in the pupae although the chlorfenvinphos seedbed treatment and the diazinon field treatment appeared to have done so.

2.1 Addendum
Plants grown in peat blocks

The control of cabbage root fly in the field, by pesticide applied only in the seedbed, is effected by the impregnation of plant roots with pesticide. The effectiveness of treatment is reduced, therefore, if the interval between treatment and transplanting is too short to allow adequate impregnation of the roots. The recommended application rate for peat block plants is about six times as great as the effective rate for seedbed-grown plants. It was postulated that insecticide, applied to peat blocks immediately before transplanting, will not impregnate roots immediately and may never do so if the insecticide remains absorbed in the peat, for example in dry field conditions. Control will, therefore, depend on contact between larvae and available insecticide in the soil, which will also be reduced in dry conditions. Earlier treatment, followed by normal watering of the plants, may render insecticide more freely available in the blocks and allow greater impregnation of the roots before transfer to the field. Martin and Mowat (1) confirmed that the performance of chlorpyrifos was enhanced by earlier application, so that the application rate could be reduced. The rate remained higher than for seedbed treatment, and the performance of chlorfenvinphos was not similarly enhanced.

I received results of two peat-block experiments from Saillon, Switzerland. In the first, treatments (Table III) were applied

post-emergence on 23 March 1983, 30 days before transplanting. The
treatments were equivalent to the lower rate suggested for seedbed
treatment. All increased yield substantially (P<0.01) compared with
untreated plots and (with the exception of the chlorpyrifos granule
treatment) were not different from a standard chlorfenvinphos field
treatment, although the latter gave greater reductions of root damage.
Increased application rates in the second experiment, treated with
insecticide on 28 March 1984 (27 days before transplanting) did not
noticeably increase the level of control, except in the case of
chlorpyrifos granules. Although root damage (Table III) indicated a
level of attack similar to the previous year, yields did not differ
significantly between treatments; and the post-emergence treatments
performed similarly (in relation to the untreated and standard
treatments) to the first experiment. Carbofuran granules were slightly
phytotoxic, although treated plants yielded well.

It was concluded that, with early application, peat block treatment
was effective at low rates although ultimately allowing higher levels
of root damage then the standard treatment. The implications of
treatment date seemed, therefore, to be similar to those observed in
seedbed treatment.

Acknowledgements
This paper includes data submitted by T H Coaker, J Freuler and
A L Percivall. I am grateful to them, and to all those who helped them
and me. I also thank P Basset, B Bromand, E Brunel and R Dunne for
their co-operation.

References
1. MARTIN, S.J. and MOWAT, D.J. (1984). The effect of pre-planting
    insecticide treatments on cabbage root fly damage to cauliflower
    grown in peat blocks. Record of Agricultural Research,
    Department of Agriculture for Northern Ireland 32, 99-101.
2. MOWAT, D.J. (1970). Experiments on the control of cabbage root
    fly Erioischia brassicae (Bouché), in transplanted brassicas by
    seeded treatment. Horticultural Research 10, 142-147.
3. MOWAT, D.J. (1971). A method for the control of cabbage root fly,
    Erioischia brassicae (Bouché), in transplanted brassicas by
    seedbed treatment. Horticultural Research 11, 98-106.
4. MOWAT, D.J. (1975). The influence of application precision and
    rainfall on cabbage root fly control by field-applied
    insecticides. Record of Agricultural Research, Department of
    Agriculture for Northern Ireland 23, 19-22.
5. MOWAT, D.J. and MARTIN, S.J. (1980). A comparison of insecticides
    applied to a brassica seedbed for the post-transplanting control
    of cabbage root fly, Delia brassicae (Wied.). Record of
    Agricultural Research, Department of Agriculture for Northern
    Ireland 28, 37-40.
6. MOWAT, D.J. and MARTIN, S.J. (1981). Seedbed treatments for the
    control of damage by cabbage root fly, Delia brassicae (Wied.), to
    transplanted summer cauliflowers. Horticultural Research 21,
    113-125.
7. MOWAT, D.J. and MARTIN, S.J. (1981). The contribution of
    predatory beetles (Coleoptera: Carabidae and Staphylinidae) and
    seedbed-applied insecticide to the control of cabbage root fly,
    Delia brassicae (Wied), in transplanted cauliflowers.
    Horticultural Research 21, 127-136

The application of insecticides for cabbage root fly (*Delia radicum*) control in cabbage grown in small peat modules

R.Dunne & J.Coffey
An Foras Taluntais, Kinsealy Research Centre, Dublin, Ireland

Summary

The application of insecticides appropriate for cabbage root fly control in brassicas was investigated in cabbage grown in small peat modules. Incorporation of chemicals in the compost was found to lead to an unacceptable level of phytotoxicity. Drench application did not cause phytotoxic effects but the level of field control of the pest was less than with conventional application methods.

1.1 Introduction

Reduction of the insecticide application rate for cabbage root fly control can be achieved by treatment of brassica plants grown in peat modules either by incorporation in the compost or by preplanting insecticidal drenches. While appropriate insecticides can be safely applied to large peat modules (37 to 50 mm^3) at dosages that will give adequate field control, there are no recommendations for their application to small modules e.g. Hassy trays. (Maude and Thompson, 1983). Experiments described were designed to explore the feasibility of insecticidal treatment of plants grown in such small modules.

1.2 Experimental

Cabbage cv. Greyhound was grown in Hassy trays, 3.1 cm size (refers to spacing between plants), containing 308 cells /tray. Each cell is 5.8 cm^2 tapering to 2 cm^2, and 4 cm deep, containing approximately 15 ml compost. Drenches were applied in 1 l water per tray. Incorporated insecticides were mixed manually with the compost. Seed was also sown manually. The root damage index (R.D.I.) was calculated according to Coaker and Wheatley (1971). Seedling assessments were made when plants had reached the transplanting stage. Insecticide application rates are given as mg active ingredient per plant.

1.3 Results

The results of incorporating the insecticides in the compost are shown in table 1.

The 20 mg rates of all insecticides and the 10 mg level of chlorfenvinphos resulted in a degree of damage that made the plants unsuitable for transplanting. All treatments except diazinon at the 5 mg level inhibited root development resulting in fragmentation of modules at planting in addition to malformation and stunting of foliage. In general, insecticide incorporation resulted in unacceptable damage to plants by all

Table I

Incorporation of insecticides in compost

Treatments			Yield kg/plot	No. mkt. hds.	R.D.I.
Chlorpyrifos 5% G	5	mg	22.5**	26.3***	5.3
	10	"	16.5	23.5***	1.9**
	20	"		not transplanted	
Chlorfenvinphos 10% G	5	"	18.3*	27.0***	9.2
	10	"		not transplanted	
	20	"		not transplanted	
Diazinon 5% G	5	"	18.5*	24.5***	13.2
	10	"	18.1*	22.5**	16.6
	20	"		not transplanted	
CONTROL			11.9	15.5	10.6
S.E.			1.8	1.4	2.3

Table II

Drench treatments to Hassy trays

Treatments		No. of seedlings	wt. of seedlings (g)
Chlorpyrifos	0.5 mg	99.7	56.7
48% e.c.	1.9 mg	104.0	64.0
	2.0 mg	107.3	66.0
	4.0 mg	109.7	84.0**
CONTROL		105.3	65.3
S.E.		4.0	4.0

Table III

Drench treatments to Hassy trays

		Seedling vigour 1 - 10	Weight of seedlings(g)
Chlorpyrifos	1 mg	8.6	303.0
48% e.c.	2 mg	8.4	342.3
	3 mg	8.4	349.0
	4 mg	8.0	368.0
	5 mg	8.9	390.5*
S.E.		0.3	23.2

the insecticides used and it was concluded that the incorporation of a
field effective dosage in such small modules was impracticable due to the
phytotoxic effects of the insecticides on the plants. Consequently, it was
thought that topical application (drenches) of the insecticides to plants
immediately before transplanting presented the better chance of successful
control without phytotoxic effects. Chlorpyrifos was considered the least
phytotoxic insecticide with known effectiveness against the pest.

Table $\overline{IV}$

Field performance of drench treated plants (1985)

		Yield(kg)	Mkt. heads	R.D.I.
Chlorpyrifos	1 mg	15.9*	24.8***	32.7***
48%	2 mg	16.0*	24.3***	27.3***
	3 mg	17.2*	24.0***	29.1***
	4 mg	14.9	21.5*	22.4***
	5 mg	15.9	22.5**	20.5***
CONTROL		11.7	18.5	51.6
S.E.		1.4	0.9	3.1

Table $\overline{v}$

Field performance of treated plants (1985)

			Yield (kg)	Mkt. heads	R.D.I.
Chlorpyrifos	48% e.c.	5 mg	22.4***	24.3	28.3***
"		10 mg	20.8**	25.5*	28.1***
Diflubenzuron	25% w.p.	5 mg	17.7	20.0	54.6
"		10 mg	21.3***	23.8	42.7**
"		20 mg	20.8**	24.3	37.7***
Cyromazine	75% w.p.	5 mg	16.0	19.8	47.6*
"		10 mg	15.5	18.0	44.4**
"		20 mg	15.4	18.5	36.6***
Fonofos	43% m.s.	5 mg	23.8***	24.0	26.4***
"		10 mg	21.4***	22.8	34.9***
CONTROL			15.7	20.3	59.7
S.E.			1.1	1.6	3.4

Results indicated that chlorpyrifos at rates of 4 mg and lower did not affect seedling establishment and seedling vigour - in fact there were indications that the insecticide had a slightly positive effect on seedling growth. This is also indicated by results in table $\overline{III}$.

Table $\overline{IV}$ shows that while all rates of chlorpyrifos significantly increased the number of marketable heads and reduced the root damage index this was not consistently reflected in the total yield. Results also indicate that while the standard application rate of chlorpyrifos to large modules is 10 mg per plant, rates in the range 1-5 mg gave a significant reduction in root damage which, particularly in the case of cabbage, may be sufficient to give an acceptable level of control.

Table $\overline{V}$ shows that all treatments except diflubenzuron at 5 mg and all rates of cyromazine significantly increased yield. However, while the reduction in root damage was significant in all treatments except diflubenzuron at 5 mg, and this reduction was reflected in yield increase, the resulting levels of attack are probably unacceptable in practice. As[2] the chlorpyrifos and fonofos treatments did not give a higher level of control, it is suggested that drenching such small modules does not result in an adequate field dosage being carried by the peat in the modules.

However, this conclusion may need to be revised in the context of an integrated pest management approach.

1.4 Conclusions

It is desirable that recommendations are available for the use of insecticides to protect brassica plants grown in small peat modules against field populations of cabbage root fly. Attempts to investigate the feasibility of adding adequate amounts of appropriate insecticides to the compost resulted in an unacceptable level of phytotoxicity to plants. Topical application of insecticides by drenching plants immediately before transplanting did not lead to significant phytotoxic effects but the degree of control achieved in the field, while frequently reflected in yield increases, was less than expected from conventional field treatments.

1.5 References

Coaker, T.H. and Wheatley, G.A. (1971). Crop Loss Assessment Methods, FAO Manual.

Maude, R.B. and Thompson, A.R. (1983). Pests and Diseases of Vegetables. Pest and Disease Control Handbook, Chapter 10, Ed. N. Scopes and M. Ledieu, BCPC Publications.

Investigations of the resistance of cabbage cultivars and breeders lines to insect pests at Wellesbourne

P.R.Ellis & J.A.Hardman
National Vegetable Research Station, Wellesbourne, Warwick, UK

Summary

The resistance of a wide range of cabbage material to several important insect pests was assessed in field experiments in five seasons at Wellesbourne. None of the 70 cultivars and breeders lines tested were found to possess resistance to all pests. Both non-preference for egg-laying and resistance to maggot damage were identified with respect to cabbage root fly (Delia radicum) attack; three red cabbage cultivars and two USA lines were the least damaged by maggots. There were no significant differences between 11 cultivars and lines in the number of plants infested by cabbage stem weevil (Ceutorhynchus pallidactylus).

None of the cabbages was resistant to flea beetle (Phyllotreta spp.) attack and severity of damage was highly correlated with glossiness of plant foliage.

The apparent resistance of one green and two red cabbage cultivars to cabbage aphid (Brevicoryne brassicae) early in the season changed as the plants aged; however, two US lines were consistently less damaged by aphids. High levels of resistance to lepidopterous pests bred into certain USA lines were confirmed for small white butterfly (Pieris rapae), cabbage moth (Mamestra brassicae), diamond-back moth (Plutella xylostella) and garden pebble moth (Evergestis forficalis). Two of these lines also possessed resistance to cabbage root fly and cabbage aphid.

1. INTRODUCTION

The use of insect-resistant cultivars offers an attractive method of supplementing insecticides to achieve a satisfactory control of pests of the cabbage crop, Brassica oleracea L. var. capitata (L.) Alep. The search for resistance in cabbage to certain important pests has therefore been the objective of research at several centres in the world, the investigations at Wellesbourne beginning in 1971. There are numerous accounts in the literature of resistance to the more important pests, the earliest report for cabbage root fly, Delia radicum (L.) dating back nearly 70 years. Gibson and Treherne (9) discovered that certain red cabbage cultivars were less damaged than several others in field experiments. Brittain (3) compared 22 cultivars and found 'Large Red Drumhead' to be totally immune to maggot attack, the resistance of this Savoy type cabbage being confirmed in later experiments (4). These reports were followed by several others (for example 10, 22, 25) including three which confirmed the resistance of red cultivars (16, 17, 20). Maack (15) showed that the growth stage of cabbage cultivars greatly influenced root fly egg-laying. He made use of this information in devising an integrated programme of cabbage pest control which included choice of cultivar and sowing time to avoid the worst insect damage.

Investigations on foliage pests of cabbage date back nearly 40 years when differences were observed between cultivars in their resistance to cabbage aphid Brevicoryne brassicae (L.), and three lepidopterous pests (1, 11). In a series of experiments between 1958 and 1962 Radcliffe and Chapman (18, 19, 20, 21) studied the resistance of 21 cabbage cultivars to cabbage aphid, and three lepidopterous pests as well as cabbage root fly. Seasonal shifts in the relative resistance of the cultivars were observed with respect to the different pests and no single cultivar was resistant to all species. Brett and Sullivan (2) summarised a series of experiments carried out over a seven-year period on pests of a range of cruciferous crops including 37 cabbage cultivars. High levels of resistance were found to eight different insect species in various cabbage cultivars although no single cultivar possessed resistance to all pests. Lara and co-workers found that cabbage aphid had a preference for certain cabbage cultivars and two of these also possessed antibiosis resistance (13, 14); leaf age was found to influence the levels of resistance to aphids. Dickson and Eckenrode (6) reported that high levels of resistance to the cabbage looper (Trichoplusia ni (Hubner), small white butterfly, Pieris rapae (L.) and the diamond-back moth, Plutella xylostella (L.) were transferred from a cauliflower line, PI 234599 to cabbage. The resistance was maintained irrespective of plant age or the presence or absence of alternative hosts of the pests. Inconsistent results in field experiments with cabbage aphid and the large white butterfly, Pieris brassicae (L.), were reported by Verma et al. (24) who screened 47 cabbage cultivars for resistance. Resistance to aphids was not correlated with resistance to the caterpillars. Hommes (12) studied the integrated control of pests of cabbage in Germany and of 23 insect species recorded only five caused serious damage. Resistance in certain cabbage cultivars was found to complement insecticide use and it was possible to reduce the number of spray applications on the partially-resistant cultivars.

In this paper we describe the results of a series of experiments on a search for resistance in cabbage to various pests carried out at Wellesbourne.

2. MATERIALS AND METHODS

All experiments were carried out in a sandy loam soil in fields at Wellesbourne. Details of the cabbage cultivars tested, transplanting dates, number of plants examined and insect species recorded are given in Table 1. Cabbage plants were raised in 4.3 x 4.3 cm peat blocks in trays in a glasshouse and later in frames. The cabbage plants were transplanted in the field in rows 60 cm apart, spaced 45 cm within rows and arranged in randomised blocks. The number of cabbages within a block and the number of blocks (three to eight) in an experiment depended on the availability of plants. No insecticides or fungicides were used except for cabbage root fly control in the 1972 experiment when chlorfenvinphos (Birlane Granules; Shell) was applied at a rate of 0.17 g/plant to the last two plants in each row and in 1983 when the first five plants in one block were treated with chlorfenvinphos at the same rate.

To determine the numbers of root fly eggs laid around cabbages in 1971, the soil surrounding each test plant was removed to a depth of 3 cm and to within 5 cm diameter of the stem with a spoon and the soil transferred to a labelled container. Fresh soil was added around the plants to restore the site. Soil sampling continued at two- or three-day intervals during the period of egg-laying activity of the flies. Eggs were recovered

TABLE 1 <u>Details</u> <u>of</u> <u>the</u> <u>Series</u> <u>of</u> <u>Field</u> <u>Experiments</u> <u>at</u> <u>Wellesbourne</u> <u>Between</u> <u>1971</u> <u>and</u> <u>1985</u> <u>to</u> <u>Investigate</u> <u>the</u> <u>Resistance</u> <u>of</u> <u>Cabbage</u> <u>Cultivars</u> <u>to</u> <u>Insect</u> <u>Pests</u>

Year	No. of cultivars	No. of plants examined per cv.	Transplanting date	Insect species recorded							
				Delia radicum	_Brevicoryne brassicae_	_Pieris rapae_	_Mamestra brassicae_	_Plutella xylestella_	_Evergestis forficalis_	_Phyllotreta_ spp.	_Ceutorhynchus pallidactylus_
1971	8	12	23 July	X							
1972	12	190	25 April	X							
1973	12	160	30 April	X							
1983a	50	60	4 May	X	X	X		X			
1983b	50	60	21 June		X	X		X	X		
1984a	12	120	10 May	X	X	X	X	X	X	X	X
1984b	12	120	27 June	X	X	X	X	X	X		

from the soil using a flotation system devised by Ellis and Hardman (8).

Maggots damage to the root system of cabbage plants was assessed using a standard root damage index (23). Plants were lifted, shaken free of soil, weighed and assigned to one of four damage grades according to the severity of attack. Mean damage grades were calculated for each cultivar.

A grading system was also used for recording numbers of insects and the damage resulting from cabbage aphid, caterpillar and flea beetle (<u>Phyllotreta</u> spp.,) attack:- 1) no damage present; 2) slight damage; 3) moderate (up to 50% leaves damaged); and 4) severe damage (50% of leaves damaged). In situations where cabbage aphid colony size could readily be estimated, numbers of insects were recorded using the grading system devised by Dunn & Kempton (personal communication):- 1) no aphids present;   2) 50 aphids scattered and with no colony development;   3) up to three colonies no greater than 1 cm diameter; and 4) three colonies.

In 1984 the presence or absence of cabbage stem weevil, Ceutorhynchus pallidactylus (Marsh.), damage was recorded.

3. RESULTS AND DISCUSSION

3.1   Cabbage Root Fly (Delia radicum)

There was considerable variation in the numbers of eggs laid around individual plants within a cultivar and between occasions during the egg-laying period in the 1971 experiment. Despite the variation, there were significant differences between the cultivars in the flies' preferences for egg-laying, the two least preferred cultivars being 'Golden Acre' and 'Persista'. There was also some indication of a relationship between egg-laying and plant size, the largest plants being more preferred. Cv. 'Golden Acre' was found to be intermediate in its attractiveness to root fly egg-laying amongst 21 cabbage cultivars screened by Radcliffe and Chapman (20, 21). Egg-laying preferences were not recorded in subsequent experiments at Wellesbourne, partly because of the enormous variation within cultivars and during the egg-laying period, and because other workers had indicated that a cultivar which was most preferred for egg-laying could be the least damaged by the maggots (20). Ellis and Hardman (unpublished) observed a similar phenomenon when investigating the resistance of radish to cabbage root fly attack. In addition, under 'no choice' conditions similar numbers of eggs may be laid on different cabbage cultivars (5). The same eight cultivars as tested in 1971 were included in investigations of the tolerance of cabbage cultivars to cabbage root fly attack in 1972 and 1973. In both years plants established rapidly. Cabbage plants treated with insecticide in 1972 received a check in growth initially but eventually produced more vigorous plants than those in untreated plots. In both years root fly damage was slight, only 16% of plants showing symptoms of damage on the most severely attacked cultivar, 'Avon Coronet'. The light attack on the cabbages contrasted markedly with the heavy attack on cauliflower, Brassica oleracea L. var. gemmifera Zenk., in neighbouring plots in both experiments. The cultivar 'Persista' which had been the least preferred cultivar for egg-laying was also the least damaged by maggots.

Fifty cabbage cultivars including 27 landraces from Ireland and several cultivars from the USA which were reported to be resistant to pests, were tested in 1983. Cabbages were grown as two plantings in order that they could be screened against a wide range of pests in the spring and summer months. It was possible to identify a few cultivars which were resistant to several pests. Five cultivars, 'Red Drumhead', 'Super Red', 'Drumhead Late', 'Late Purple Flat Poll' and 'Yates Giant Red' were significantly less damaged by cabbage root fly than the other seedstocks tested. None of the landraces were resistant to root fly damage. Seven of the cultivars tested in 1983, including the most promising cultivars listed above, were evaluated in a field experiment in 1984 and grown alongside cabbage material obtained from the USA which had formed part of a breeding programme aimed at developing insect-resistant cabbage cultivars (6). Cabbage root fly damage was severe on the first transplanting while the second largely escaped attack. Two US breeding lines G 10127 and G 10128 together with cvs 'Little Rock' and 'Super Red' were the least damaged, the results for 'Super Red' confirmed those obtained in 1983.

3.2 Cabbage Aphid (Brevicoryne brassicae)

Two red cabbage cultivars, 'Red Drumhead' and 'Yates Giant Red', were among the least attacked of 50 cultivars tested in the spring of 1983 but they became less resistant to aphid infestation as the season progressed. Similar seasonal changes in the relative resistance of cabbage and related crops to cabbage aphid have been observed by other workers (7, 20). In 1984 seven of the cultivars included in the 1983 experiment were tested alongside five seedstocks from the USA (see earlier). The seasonal shift in the resistance of cv. 'Red Drumhead' to cabbage aphid was confirmed while two breeding lines G 10127 and G 10128 from the USA were consistently less damaged than other cultivars and, for these, no shift in relative resistance was observed during the season.

3.3 Cabbage Caterpillars (Pieris rapae, Mamestra brassicae, Plutella xylostella and Evergestis forficalis)

Caterpillar damage was recorded on both transplantings of the 50 cabbage cultivars in 1983. Infestations of diamond-back moth and small white butterfly built up in July when plants of some cultivars were 15 weeks old and had produced marketable-sized heads. The same two species of lepidoptera also infested the second transplanting in late July when plants were six weeks old. By late August garden pebble moth caterpillars were present in the second transplanting in large numbers. Very few cabbage moth caterpillars were found in this experiment. Records showed that several landraces from Ireland were very severely damaged while cvs 'Red Drumhead', 'Yates Giant Red', 'Super Red' and 'Resistant Danish' were only slightly damaged by all three insect species in both transplantings.

The cultivars 'Red Drumhead' and 'Super Red' were included in the 1984 experiment and compared with ten other lines including those bred for resistance to caterpillars in the USA. Records of caterpillar damage were collected for both transplantings in this experiment, four lepidopterous species being present in significant numbers. The two cultivars 'Red Drumhead' and 'Super Red' were promising, thus confirming the 1983 results. However, they were less resistant than two USA cabbage breeding lines, G 10127 and G 10128, which were practically free of damage in both transplantings. Two other lines, G 10114 and G 10131, were as resistant as the two red cabbage cultivars.

3.4 Flea Beetle (Phyllotreta spp.)

In 1984 flea beetle damage was recorded on every plant in the first transplanting of cabbages. At the time of recording damage, plants were also scored on the glossiness of their foliage. None of the cultivars was undamaged and severity of damage was highly correlated with glossiness of the leaves.

3.5 Cabbage Stem Weevil (Ceutorhynchus pallidactylus)

None of the cabbage cultivars and breeders lines tested in 1984 were resistant to stem weevil. The least damaged cv. was 'Super Red' with 70% of plants infested while all other cvs had more than 85% plants damaged.

4. CONCLUSIONS

The cabbage cultivars and lines screened represented a wide range of cabbage genotypes and also represented the range of responses to pests as reported in the literature. No single cultivar was resistant to all insect species although cv. 'Super Red' and two USA breeding lines were clearly promising when tested against several important pests and maintained their resistance throughout the period between May and early September. Although the levels of resistance were only moderate to certain pests they may still make a valuable contribution to pest control when used in collaboration with insecticides. Their contribution to pest control could be particularly significant in cases such as with cabbage root fly infestations where the marketed produce is usually unattacked by the insect.

REFERENCES

(1)   BREAKEY, E.P. and CARLSON, E.C., 1944. Control of the cabbage aphid (Brevicoryne brassicae) in cabbage plant beds and seed fields. Washington Agricultural Experiment Station Bulletin No. 455, 38 pp.

(2)   BRETT, C.H. and SULLIVAN, M.J., 1974. The use of resistant varieties and other cultural practices for control of insects on crucifers in North Carolina. North Carolina State University Agricultural Experiment Station Bulletin No. 449, 31 pp.

(3)   BRITTAIN, W.H., 1922. Further experiments in the control of the cabbage maggot (Chortophila brassicae Bouche) in 1921. Proceedings of the Acadian Entomological Society for 1921, pp. 49-71.

(4)   BRITTAIN, W.H., 1927. The cabbage maggot and its control in Nova Scotia. Nova Scotia Department of Natural Sciences Bulletin No. 11, 53 pp.

(5)   CRUGER, G. and MAACK, G., 1980. Using an economic threshold to reduce the amounts of insecticides applied to control H. brassica. IOBC-WPRS Bulletin 1980/III/I. pp. 27-34.

(6)   DICKSON, M.H. and ECKENRODE, C.J., 1980. Breeding for resistance in cabbage and cauliflower to cabbage looper, imported cabbage worm, and diamond-back moth. Journal of the American Society of Horticultural Science 10, 782-785.

(7)   DUNN, J.A., 1978. Resistance to some insect pests in crop plants. In: Applied Biology Vol. III (Ed. Coaker, T.H.). London: Academic Press, pp. 43-85.

(8)   ELLIS, P.R. and HARDMAN, J.A., 1975. Laboratory methods for studying non-preference resistance to cabbage root fly in cruciferous crops. Annals of Applied Biology 79, 253-264.

(9)   GIBSON, A. and TREHERNE, R.C., 1916. The cabbage maggot and its control in Canada, with notes on the imported onion maggot and the seed corn maggot. Canadian Department of Agriculture Entomology Branch Bulletin No. 12, 58 pp.

(10) GLASGOW, H., 1925. Control of the cabbage maggot in the seedbed. New York State Agricultural Experiment Station Bulletin No. 512, pp. 5-59.

(11) HARRISON, P.K. and BRUBAKER, R.W., 1943. The relative abundance of cabbage caterpillars on cole crops grown under similar conditions. Journal of Economic Entomology 36, 589-592.

(12) HOMMES, M., 1983. Untersuchungen zur Populations-dynamik und integrierten Bekampfung von Kohlschädlingen. Mitteilungen aus der Biologischen Bundesanstalt für Land- und Forstwirtschaft, Berlin-Dahlem Heft 213, 210 pp.

(13) LARA, F.M., COEHLO, A. and MAYOR, J. 1979. Resistencia de variedades de couve a *Brevicoryne brasicae* (Linnaeus, 1758). II. Antibiose. Anais da Sociedade Entomologica do Brasil 8, 217-223.

(14) LARA, F.M., MAYOR, J., COEHLO, A. and FORNASIER, J.B., 1978. Resistencia de variedades de couve a *Brevicoryne brassicae* (Linnaeus, 1758). I. Preferencia em condicoes de campo e laboratorio. Anais da Sociedade Entomologica do Brasil 7, 175-182.

(15) MAACK, G., 1977. Schadwirkung der Kleinen Kohlfliege (*Phorbia brassicae* Bouche) und Möglichkeiten zur Reduzierung des Insektizidaufwandes bei der Bekämpfung. Mitteilungen aus der Biologischen Bundesanstalt für Land- und Forstwirtschaft, Berlin-Dahlem Heft 177, 135 pp.

(16) MATTHEWMAN, W.G. and LYALL, L.H., 1966. Resistance in cabbage to the cabbage maggot, *Hylemya brassicae* (Bouche). Canadian Entomologist 98, 59-69.

(17) NIKITINA, T.F., 1938. The biology and ecology of *Hylemya brassicae* Bouche in Gorki Province. Plant Protection No. 17, pp. 79-85.

(18) RADCLIFFE, E.B. and CHAPMAN, R.K., 1965. Seasonal shifts in the relative resistance to insect attack of eight commercial cabbage varieties. Annals of the Entomological Society of America 58, 892-897.

(19) RADCLIFFE, E.B. and CHAPMAN, R.K., 1965. The relative resistance to insect attack of three cabbage varieties at different stages of plant maturity. Annals of the Entomological Society of America 58, 897-902.

(20) RADCLIFFE, E.B. and CHAPMAN, R.K., 1966. Plant resistance to insect attack in commercial cabbage varieties. Journal of Economic Entomology 59, 116-120.

(21) RADCLIFFE, E.B. and CHAPMAN, R.K., 1966. Varietal resistance to insect attack in various cruciferous crops. Journal of Economic Entomology 59, 120-125.

(22) SCHOENE, W.J., 1916. The cabbage maggot: its biology and control. New York State Agricultural Experiment Station Bulletin No. 419, pp. 99-160.

(23) THOMPSON, A.R., 1980. The assessment of damage caused to cruciferous crops by larvae of *Hylemya brassicae* in soil. IOBC-WPRS Bulletin 1980/III/I, pp. 19-26.

(24) VERMA, T.S., BHAGCHANDANI, P.M., NARENDRA SINGH and LAL, O.P., 1981. Screening of cabbage germplasm collections for resistance to *Brevicoryne brassicae* and *Pieris brassicae*. Indian Journal of Agricultural Science 51, 302-305.

(25) WHITCOMB, W.D., 1944. The cabbage maggot. Massachusetts Agricultural Experiment Station Bulletin No. 412, 28 pp.

Sequential sampling of insect pests in Brussels sprouts

J.Theunissen
Research Institute for Plant Protection, Wageningen, Netherlands

Summary

Sequential sampling charts have been developed for use in Brussels sprouts. Based on variable tolerance levels for the caterpillar complex and cabbage aphid infestation levels during the entire growing period appropriate sequential sampling schemes for both categories of pests are given. Their value as a tool for the grower under various conditions is discussed. Sequential sampling seems to be useful at tolerance levels of 20 % of infested plants and higher. Sampling takes increasingly less time when infestation differs more from the relevant tolerance levels. Varietal differences may influence sampling costs.

1. Introduction

Cabbage crops are among the economically most important vegetables all over the world. In the temperate regions these crops are infested by a varying number of insect pest species. Despite this variety most pest species can be grouped in complexes such as the caterpillar complex and the aphid complex. When possible it is practical to direct control measures against such complexes in stead of against individual species. This enhances the simplicity of pest management procedures where simplicity is needed to enable growers to take pest control decisions by themselves. Supervised control methods as a starting point of pest management in cabbage crops have been developed. Depending on the local conditions the target groups of practical implementation of these methods are different. In the United States methods have been developed for use by scouts and pest management consultants (Hoy et al., 1983), whereas in Europe the grower himself is supposed to be able to carry out the field sampling and to take subsequently pest control decisions (Theunissen & den Ouden, 1985).
In supervised control two tools are needed:
1. a simple and practical field sampling recipe
2. a set of criteria for the sampling result: fixed (Wilson et al., 1983), or variable tolerance levels (Theunissen, 1984; Theunissen & den Ouden, 1985) as practical approximations of theoretical control thresholds.
A field sampling recipe for cabbage crops based on systematic sampling has been developed (Theunissen & den Ouden, 1983; Theunissen, 1984a) and sets of variable tolerance levels have been tested for a number of years in commercial growing of Brussels sprouts and white cabbage (Theunissen & den Ouden, 1985). While the above mentioned field sampling method is functioning satisfactorily for heading cabbage crops, the growth of the Brussels sprouts plants during the season causes an increasing time spent on samp-

ling. This disadvantage of the introduced sampling method encouraged growers to devise their own variants with considerably reduced sample sizes. To
prevent possible failures of supervised control due to wrong sampling less
time consuming sampling methods for use in Brussels sprouts have been developed by means of sequential sampling.

2. Methods

Since the infestation of the crop is expressed in percentage infested
plants as are the variable tolerance levels sequential sampling schemes
have been based on a binomial distribution of infested plants. The desired
precision has been set at $\pm$ 20 % of the relevant tolerance level. For
type I errors, overestimation of the pest population, the acceptable probability α has been fixed at 0.30. The acceptable probability β of type II
errors, underestimation of the pest population, has been determined at a
lower level of 0.05 (see discussion). The calculation of these schemes took
place according to Onsanger (1976).
The characteristics of the sequential sampling schemes for the various
tolerance levels are presented in Table I.

Table I. Data of sequential sampling charts for use in supervised control
of pests in Brussels sprouts, based on a precision of $\pm$ 20 %,
α = 0.30 and β = 0.05.

tolerance level*	intercept of H1	H2	slope
4	−6.251	2.731	0.0394
10	−5.865	2.563	0.0988
20	−5.220	2.280	0.1980
40	−3.918	1.711	0.3983
50	−3.254	1.421	0.5001

* in % infested plants

The charts consist of regression lines (H) expression the relationship
between accumulated sample size (x-axis) and the accumulated number of infested plants (y-axis) under the given conditions as they relate to crop
protection decisions. The regression lines divide the quadrant between both
axes in three sections: a zone of indecision between the straight lines
which indicates a need for more samples, a section below the lowest line
which represents the decision that a treatment is not necessary, and the
top section which stands for the decision that a control treatment is
necessary.

3. Results

In Fig. 1 the sequential sampling charts for the various tolerance levels of Brussels sprouts are given.

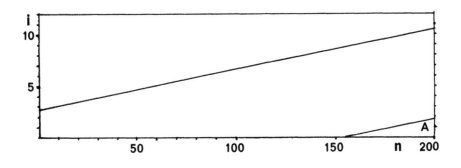

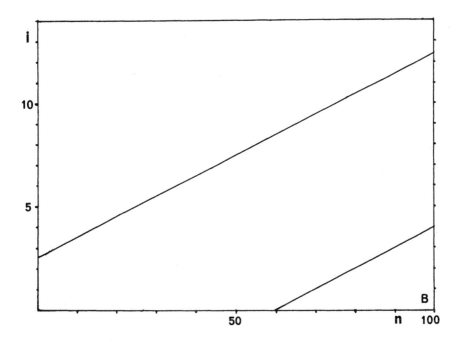

Fig. 1 A-E: Sequential sampling charts for use in Brussels sprouts.
On the abcissa is the accumulated number of samples (n),
on the ordinate the accumulated number of infested plants
(i) mentioned. The charts are valid for the following
tolerance levels: A = 4 %, B = 10 %, C = 20 %, D = 40 %
and E = 50 % infested plants.

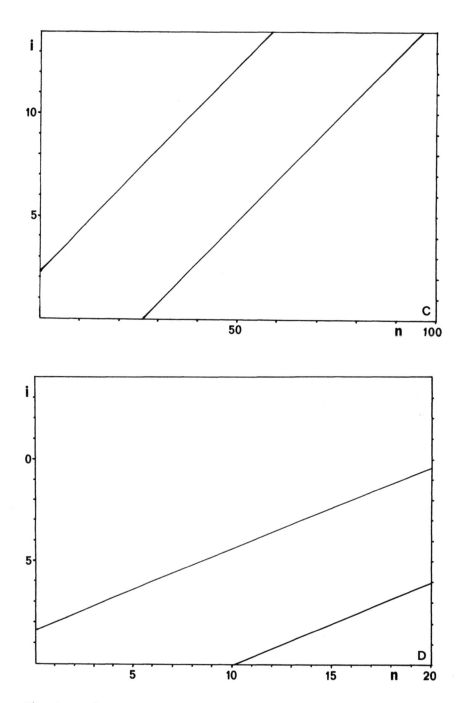

Fig. 1 continued

110

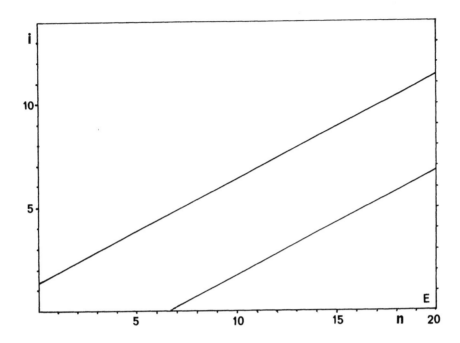

Fig. 1 continued

To facilitate the use of sequential sampling in the field a tabular
form of these charts is presented in Table II.

4. Discussion

Successful implementation of pest management methods, including supervised control, which are to be followed by the growers requires a maximum
of simplicity and a minimum of time. Simplicity is reached by developing as
much as possible tolerance levels for pest complexes. Under Dutch conditions the caterpillar complex in cabbage consists of the larvae of Mamestra
brassicae, Pieris rapae, Plutella xylostella and Evergestis forficalis.
Other species of Lepidoptera may occur in cabbage crops but are not considered regular pests for various reasons. The aphid complex consists under
the prevailing conditions of one species only, Brevicoryne brassicae. Pests
of local importance are Delia brassicae and Contarinia nasturtii. The
latter impairs the full implementation of supervised control in white
cabbage in an important growing region. Differences in sensitivity of the
crops for infestation are reflected in variable tolerance levels and expressed in simple terms: percentage of infested plants. Growth stage indications refer to weeks after transplanting for maximum simplicity, although
more refined definitions are available (Theunissen and Sins, 1984).
The time spent while following the prescribed method in the field is
very important for a widespread acceptance by the growers. Therefore, sampling time is crucial in practical and scientifically justified implementation. For low crops such as the heading cabbage types the recommended sampling method is quite satisfactory, especially in situations in which the

Table II: Sequential sampling tables for caterpillars and cabbage
aphids in cabbage crops.

	NO TREATMENT	number of:	TREATMENT
tolerance level (%)	infested plants	sampled plants	infested plants
4		20	4
		50	5
		80	6
		100	7
		160	10
	1	200	11
10		20	5
		30	6
		40	7
		50	8
	0	60	9
	1	70	10
	2	80	
	3	90	
	4	100	
20		20	7
	0	30	9
	2	40	11
	4	50	13
	6	60	15
	8	70	17
	10	80	19
	12	90	21
	14	100	23
40	0	10	6
	1	13	7
	2	15	8
	3	18	9
	4	20	10
	8	30	14
	15	50	22
	35	100	42
50	0	7	5
	1	9	6
	2	11	7
	3	13	8
	4	15	9
	5	17	10
	6	19	11
	11	30	17
	21	50	27
	46	100	52

need for a control treatment is not obvious. In a situation of an extremely low or high pest population the outcome is clear very soon and continued sampling up to the fixed sample size will be hard to expect from the growers. In Brussels sprouts, a tall growing crop, sampling takes increasingly more time during the growing season. In red cabbage the maximal time spent during the growing season to sample 100 plants/ha is 64 minutes (1.06 man-hours) (Theunissen & den Ouden, 1983). In Brussels sprouts this time is 162 minutes (2.7 manhours) (Theunissen, 1984b).

To optimize sample size under various sets of conditions sequential sampling can be used. Examples have been worked out here for Brussels sprouts but the same or similar schemes may be used for other cabbage crops as well. A risk of 30 % of overestimation of the pest population has been allowed because the consequences of a wrong decision are acceptable: one or more unnecessary sprays. A smaller risk of underestimation (5 %) is taken because a wrong decision may cause economic damage to the crop when necessary sprays are delayed or omitted. The level of desired precision (± 20 %) is arbitrarily set to obtain an acceptable sampling result.

From the charts and tables it will be clear that sequential sampling is most effective in more or less extreme population conditions: a decision can be taken quickly. At populations around the tolerance levels the sampling procedure does not give a decisive answer. Based on the experience that 100 plants/ha will give a good picture of the pest population and the fact that growers in general are not prepared to take bigger samples a decision has to be taken after 100 plants have been sampled. What this decision is will be determined by local circumstances. To prevent wrong decisions due to a too small size or accidental sampling within a cluster of pests it is recommended to take at least 10 samples before making any decision and to spread the samples in the field as has been described earlier for routine systematic field sampling (Theunissen, 1984a).

Using sequential sampling will be most rewarding at the higher tolerance levels. At the 4 % tolerance level a minimum of about 160 plants will have to be sampled and found uninfested before a "no spray" decision can be taken. In this case it is obvious that the systematic sampling procedure recommended earlier is less time consuming. During the growing season of 1985 the time necessary to reach a decision has been recorded (Table III). Crops were sampled according to both the recommended systematic sampling procedure and the appropriate sequential sampling scheme at the same field at the same moment.

From these data it appears that considerable saving of sampling time may be achieved by using sequential sampling schemes if the population levels in the field differ sufficiently from the current tolerance levels. If not then sequential sampling may require a sample size even larger than the fixed sample size for routine systematic sampling, especially at low tolerance levels. Usually the tolerance levels for cabbage aphid infestation are well below those for the caterpillar complex. This means that two charts have to be used in the field which can be done without problems. Devices like a peg board or more sophisticated counting devices may be useful.

Later in the season when the Brussels sprouts plants become taller differences between varieties influence the necessary sampling time as is given in Table IV.

The difference in sampling time was caused by the habitus of the plants of both varieties. Rampart has leaves with firm stems which are placed perpendicular to the main stem. Moving between the rows is difficult and slows down the sampler. Acropolis has hanging leaves with more or less pliable stems. This leaves some space between the rows and facilitates

Table III: Time (minutes) necessary for one person to reach a decision in a crop of Brussels sprouts, when sampling for the caterpillar complex.

systematic*	sequential	% reduction**	tolerance≠	location
6***	2.5	58	50	OBS
8***	4	50	40	PAGV
8***	4.5	44	40	PAGV
55	8	85	40	PAGV
51	14	72	40	OBS
98	7	93	40	PAGV
122	19	84	40	PAGV
87	14	84	40	OBS
87	83 (1)	4	10	OBS
94	135	-44	10	PAGV
69	69	0	10	OBS

*   sample size fixed at 70 plants
**   reduction as percentage of "systematic"
***   pilot samples (n=10) at high tolerance levels and low populations densities
≠   tolerance level in % infested plants
(1)   decision taken for cabbage aphids
      location: two locations were used: the Experimental Station for Vegetable Growing (PAGV) and the Experimental Station for Development of Farming Systems (OBS), respectively at Lelystad and Nagele.
      varieties: Rampart (PAGV), Acropolis (OBS)

Table IV: Effect of varietal differences on sampling time at two tolerance levels.

location	variety	tolerance levels	
		10 %	20 %
PAGV	Rampart	135*	54
OBS	Acropolis	69	27

* sampling time in minutes by sequential sampling necessary to reach a decision

the movement of the sampler. Since most of the sampling time is spent in walking through the field these varietal differences appear to effect sampling costs. In both cases the pest situation was similar and did not cause differences in sampling time.

A problem is the required sample size for low tolerance levels. No grower is prepared to sample more than 100 plants in order to be scientifically sure that the pest population level is below a specified low level. Neither systematic nor sequential sampling offer a solution of this problem. In practice, growers will start to sample for symptoms of injury which may be old and not relevant for the present situation. Essentially the growers have the choice of spraying upon finding the first infested plant, which may be economically not feasible, or spraying at much higher tolerance levels than what they claim to accept and which are covered by the sample size they are prepared to take in the field. Based on the experience with our field verifications it appears that tolerance levels of 20 % and higher are well accounted for by using sequential sampling. The sample sizes needed here are within the range growers are prepared to take, both in number of plants and in time needed.

Acknowledgements

The assistance of Miss Anja Steenkiste and Mr Siert Folkertsma in collecting the field data is gratefully acknowledged.

REFERENCES

1. HOY, C.W., JENNISON, G., SHELTON, A.M. and ANDALORO, J.T. (1983). Variable-intensity sampling: a new technique for decision making in cabbage pest management. J. Econ. Entomol. 76: 139-143.
2. ONSAGER, J.A. (1976). The rationale of sequential sampling, with emphasis on its use in pest management. Techn. Bull. No. 1526, USDA.
3. THEUNISSEN, J. (1984a). Supervised control in cabbage crops: theory and practice. Mitt. Biol. Bundesanst. 218: 76-84.
4. THEUNISSEN, J. (1984b). Development and application of sampling methods of pests in Brassica crops. CEC Programme on Integrated and Biological Control, Final Report 1979/1983, EUR 8689: 303-307.
5. THEUNISSEN, J. and DEN OUDEN, H. (1983). Praktijkbemonstering van plagen in vollegrondsgroenten 3. Geleide bestrijding in rode kool. Gewasbescherming 14: 119-124 (in Dutch).
6. THEUNISSEN, J. and SINS, A. (1984). Growth stages of cabbage crops for crop protection purposes. Scientia Hortic. 24: 1-11.
7. THEUNISSEN, J. and DEN OUDEN, H. (1985). Tolerance levels for supervised control of insect pests in Brussels sprouts and white cabbage. Z. ang. Ent. 100: 84-87.
8. WILSON, L.T., PICKEL, C., MOUNT, R.C. and ZALOM, F.G. (1983). Presence-absence sampling for cabbage aphid and green peach aphid (Homoptera: Aphididae) on Brussels sprouts. J. Econ. Entomol. 76: 476-479.

Conditions for implementation of supervised control in commercial cabbage growing

J. Theunissen & H. den Ouden
Research Institute for Plant Protection, Wageningen, Netherlands

Summary

Based on experience with the implementation of supervised control methods in commercial cabbage growing in the Netherlands conditions and constraints influencing the success of transfer of research results into practical application are discussed.

1. Introduction

To carry out research on Pest Management methods is one thing, to get the results applied in the field by commercial growers is quite another. This is illustrated by the relatively low number of pest management methods actually used by growers in their regular cropping practices when compared to the number of pest management methods which are potentially useful or are applied by researchers on a (semi-)experimental basis.

As part of a pest management approach of insect pests in cabbage crops supervised control has been developed for the cabbage caterpillar complex and the cabbage aphid, Brevicoryne brassicae (Theunissen & den Ouden, 1985). For some years these control methods are being implemented in commercial cabbage growing in co-operation with the plant protection specialists and the horticultural extension service in the Netherlands. This is a slow and vulnerable process which requires careful consideration of the various factors which stimulate and protect it. Since we presume that this is a process which will take place elsewhere in a similar way, we will try to name, define, analyse prerequisites which are important for practical implementation of pest management under European conditions.

2. Requirements

Demands will be put forward for any new method which is recommended to the growers. These demands will originate from both extension personnel and growers and pertain to several categories.

2.1 Communication requirements

Implementation of almost everything in commercial growing depends
directly on the co-operation between researcher, extension specialists and
the progressive grower. Therefore, it is of paramount importance for the
researcher to establish and maintain good relations with those extension
workers (in the Netherlands plant protection specialists) who can contribute to the dissemination of technical and economic knowledge to the general extension worker and the growers. To do so requires early information
of these extension specialists by the researcher about interesting new
developments and an open attitude to feed useful information spontaneously
to extension. In this way they are able to anticipate adequately. It is important that extension specialists participate in co-operative experiments
in their region and form their own opinion on the feasibility of new
methods coming up from research. When they are involved in the development
of these methods in an early stage and their opinion is taken seriously
there will be no need to convert them later on. When the new method is
sound in their opinion they will defend it as their own baby.
Another condition to have a good relation between research and extension is that the researcher respects the competence and territory of the
extension worker. There should be no direct and permanent contact between
research and grower. This is the territory of extension. Contact can be
made upon mutual request and with co-operation of extension for special
reasons such as a study tour. Similarly, information from research to grower should be channelled through extension. They may adapt the information
according to their audience and opinion and they should do so because they
speak the proper language. Extension workers should be granted their own
story and the respect they derive from it from the growers. In this situation the researcher is wise to stay in the background to supply vital and
complete information to the extension specialists only. All parties concerned will benefit from such a policy.
An open attitude of the researcher is important from the point of view
of risk analysis by the extension worker. The latter has to assess somehow
to what extent the researcher is really sure of what he wants the growers
to be told. It is the extension worker who runs the risks of failure of the
recommendations given to the grower and he cannot afford to run such risks.
Therefore, early information on and participation in new developments is
very important for the mutual relation and co-operation in offering new
methods to the growers.
Researchers may write in growers' journals about new developments.
Apart from the policy to communicate real new information through extension
first virtually no grower will take this serious until the opinion of the
local extension worker has been asked. The role of such publications is to
back up and to confirm the more convincing story of the extension officer.

2.2 Technical requirements

It goes without saying that a new method or approach should be technically sound. This is a research matter. For implementation, however, it
means that the technical actions required from the grower must be feasible
from his point of view. Therefore, these technical actions should be as
simple and as short as possible. Operating water traps or light traps for
instance is out of the question. Using pheromone traps with a far more
selective catch of species and one dominating target species could be
feasible provided the species concerned can easily be recognized and the
number of traps per hectare is limited. Field sampling should be possible

in a short time and an acceptable maximum sample size is quickly reached by the grower (Theunissen, this issue). In general there are limits in what growers are willing to do in the field outside their standard growing actions.

If the benefits are perceived to be large more "complicated" and time consuming technical actions can be required. In some cases there will be an absolute minimal requirement to the growers to able to use a pest management method at all unless the extension service is able to do the job. An example is supervised control of carrot rust fly, Psila rosae. Monitoring the flies using modified Berlinger traps (den Ouden, this issue) is quite possible for a farmer who is really willing to minimize chemical control. If he wants to do it then he can use the method. Because the extension service simply does not have the necessary manpower available monitoring fly populations by extension at the farmlevel is impossible.

Proof of the technical feasibility of any method is the verification in the field under practical conditions during a number of seasons. This has been done for Brussels sprouts and white cabbage in the Netherlands (Theunissen & den Ouden, 1985). It is very important to give both extension workers and growers tangible examples of what they can expect from the technical and economic point of view. A convincing risk-analysis from such field data may lead to a feeling of security in dealing with the new methods and may enhance their adoption and implementation.

2.3 Economic requirements

One of the first questions to be asked about a new method will be concerned with the cost-benefit analysis. A minimal requirement of a new method is that the financial outcome is the same as compared to the current method provided there is some other benefit. This could be an easier decision, a faster way of achieving a certain result or not having to handle a pesticide. Most growers dislike pesticides but feel that they cannot avoid their use. When convinced of a method to skip one, more or all chemical control treatments growers are prepared to use this method at the same financial result while a minority is even prepared to accept a smaller financial return. In principle, however, the financial result should be equal or larger to trigger action. In this context costs of actions required to carry out the recommended method are seen as production costs. For instance costs of traps, sampling time and possible recording devices are considered to be added production costs. Benefits are the savings on unnecessary pesticides, machinery and spraying time. Items of environmental concern are usually not included in the economic picture. Avoided losses of yield in quantity or quality by less mechanical damage and soil compaction in the crop are mostly also not included. While minimizing additional costs savings should be maximized: a clear profit is the best incentive to be progressive.

2.4 Socio-psychological requirements

Public opinion about the pros and contras of pesticide use is changing slowly. For both the extension worker and the grower this creates a climate in which there is a certain urge to strive for a minimal use of pesticides. When this is made possible technically and economically a grower is considered to be progressive if he is willing to adopt pest management methods. Backing up extension with relevant information the researcher can make both extension officer and grower independent of biased information from pesticides salesmen. The entire set-up of this communication is to give the grower the confidence and competence to make his own pest control decisions independently from anyone, even from extension workers. To enable him to gain this decisiveness he must get both the technical tools (methods) and the moral backing up (self-confidence and knowledge of the biological rationale of the methods). Both can be developed by regional training sessions with extension workers.

Very important is the support of the growers' organization towards adoption of modern cropping methods, including pest management. In the Netherlands the growers' study clubs are important sources of information and reference in this respect. Mainly in wintertime items of interest such as choice of varieties, fertilization, crop protection, storage are point of discussion, and information from extension is reviewed and evaluated. Pest management information must be channelled to these study clubs.

3. Constraints

Given the technical feasibility of a pest management method like supervised control in cabbage crops there are a number of constraints for the implementation in commercial vegetable growing.

At the communications level considerable difficulties are possible and actually existing:

* Disrupted communications between researchers and extension workers can be a major constraint for practical implementation of research results. In this particular case communications have been established carefully and much depends on the people involved. However, in many countries the organizational set-up of both research and extension prevents smooth co-operation between workers of both sides often due to matters of competence, territory and traditional antagonism.

* The traditional cautiousness of farmers towards outsiders including researchers plays a role. This situation calls for an exclusive role of extension in their contacts with farmers as argued before.

* Overload of the growers by a stream of information. Continuous changes in many aspects of growing their crops take place. Crop protection is just one of such aspects. Growers tend to select priorities according to their personal preference; they cannot keep up with everything which is new.

* Conflichting information reaches the growers which causes uncertainty on what to believe and to do. The pesticide industry tries to obstruct the implementation of pest management methods by increasing the feelings of uncertainty of the growers. They suggest excessive risks when these methods are applied. Conservative persons within the extension service may also stick to old methods in their advices.

On the technical level constraints may be experienced by growers:

* Too complicated or time-consuming methods may form a barrier. This is a subjective matter. Even a very simple sampling method may be found to be too complicated or monitoring by means of a few pheromone traps may be considered too time-consuming. This will vary among growers but can be perceived as a real problem.

* Among cropping methods pest management methods (here supervised control) are relatively complicated. Many species of insects and fungal diseases may be present. Their identification and assessment of the population density or disease incidence is a major problem for the grower. Without very clear and simple instructions and adequate training he will not acquire the self-confidence to take pest control decisions independently.

* Growers often think that conventional chemical control leaves their crops without insects. This is not correct and populations of pest insects may be as large as those in fields under supervised control. The point is that conventionally sprayed fields are nearly never checked on effectivity of the treatment.

At the economic level constraints may be:

* The risk-aversion of the grower. Without proof that he is going to gain he usually will not be receptive for new ideas or methods. Pesticide salesmen play on this tendency of risk-aversion to obstruct adoption and implementation of pest management methods and tend to present their own advice. Sufficient and convincing field evaluation is the right remedy as advocated earlier.

* Availability of cheap insecticides. Some insecticides are very cheap for instance parathion. When such insecticides are allowed and effective relatively little can be saved by reducing the number of chemical treatments and the growers' risk-aversion may prevail. According to the Dutch governments' declared policy of supporting non-chemical and reduced chemical pest control strategies a levy on pesticides could provide added incentives for the development of such strategies.

Socio-psychological constraints have to do with attitudes and opinions within the groups of extension workers, growers and policy makers:

* Conservatism is widespread among policy makers. Usually this originates from ignorance about new developments in research and the potential benefits for practical application. It results in immobility and little support for institutes, groups and individuals working on these developments. In stead of leading by anticipating and guiding new developments they lag behind thus causing constraints for implementation.

* Extension workers also tend to be conservative because they cannot afford to give wrong recommendations. This is another, a different kind of conservatism and this changes usually once they participate in field verification. In some cases extension workers stick to old methods regardless of the testing results of the new ones.

* Besides risk-aversion there is a particular attitude of a grower which may cause a constraint for implementation of new methods. This attitude is based upon the viewing of growing crops as a kind of industrial activity in stead of working with biological phenomena and within the laws of life. In such a case a grower is open to changes provided they increase the scale of farming and permit the use of machinery, chemical compounds and laboursaving devices. Going down from the tractor to see what is happening in the crop is considered to be old-fashioned by these people who often have lost the feeling for farming as a natural process. Such an attitude is strengthened when middlemen explicitly want to buy products only when grown in a conventional way. This happens out of sheer ignorance, an unreasoned hostility towards new developments and an intuitive kind of risk-aversion of the middlemen. Increased awareness of the consumer on the possibilities of less pesticide usage could reverse this attitude.

* In the Netherlands a process of cutting back of extension services to the growers is going on. This is government policy in order to save on public services and promote free enterprise in agricultural activities. A serious constraint for introducing and implementing new methods of any kind is the sheer lack of manpower within the extension service to give the necessary training to interested growers during a sufficient period of time. What the policy makers do not perceive is that they break down also the effectivity of their agricultural research. When research results cannot be passed anymore to the commercial grower because of an ineffective extension service what is the benefit of research?

4. Conclusions

Implementing results of research in commercial vegetable growing is a long-term process which has to be guided carefully. Actively conditions must be met and constraints limited to achieve an increasing adoption and application of new ideas and methods within the farming communities.

REFERENCES

1. OUDEN, H. DEN and THEUNISSEN, J. Monitoring of the carrot rust fly Psila rosae for supervised control; this issue.
2. THEUNISSEN, J. Sequential sampling of insect pests in Brussels sprouts; this issue.
3. THEUNISSEN, J. and DEN OUDEN, H. (1985). Tolerance levels for supervised control of insect pests in Brussels sprouts and white cabbage. Z. ang. Ent. 100: 84-87.

Preliminary observations on swedes with resistance and susceptibility to turnip root fly (*Delia floralis*) in Scotland

N.Birch

Zoology Department, Scottish Crop Research Institute, Invergowrie, Dundee, UK

Summary

The biology and pest status of turnip root fly (TRF) in Scotland are reviewed, including the different damage thresholds and levels of control required for fodder and consumer swedes. The value of resistant cultivars and the background to the plant breeding project on TRF resistance and its mechanisms at SCRI are discussed, with reference to two new SCRI swede cultivars, Angus and Melfort, which have high levels of resistance. Preliminary results on oviposition, larval invasion and puparia numbers are outlined and the possible physical or chemical plant factors involved in resistance and the inferences for developing a rapid resistance screen for plant breeders are discussed. The role of host plant resistance in integrated control of TRF is briefly discussed.

1.1 Introduction

Turnip root fly (TRF, <u>Delia floralis</u> Fall.) is an important pest of swedes (<u>Brassica napus</u> L. ssp. <u>rapifera</u>) and turnips (<u>Brassica campestris</u> ssp. <u>rapifera</u>) in Scotland. Records of attack in Scotland date back to 1843 and the incidence increased sporadically in the 1930's continuing up to the present time (1,2). TRF is distributed throughout Scotland but most commonly causes economic damage in more northerly areas, on the west coast and the Isles including Orkney, Shetland and Hebrides (3). TRF was not recorded in the south-west of Scotland until 1955 (4) but now causes substantial damage in some years. TRF populations have been found from sea level up to 300 m and is more frequently a problem on lighter soils (1,3).

TRF has a host range including swede, turnip, cabbage, kale and also some cruciferous weeds, but in Scotland it is found generally attacking only swede and turnip (1). Unlike cabbage root fly (<u>Delia radicum</u> L.), a more important bi- or trivoltine pest of brassica crops in the south of Britain, TRF is generally univoltine, although there are reports of a second generation in parts of Denmark (5) and possibly Scotland (6). Adult flies start to emerge from early June onwards. In Scotland oviposition starts in late July with peak female activity in mid August, equivalent to 860 D^o above a base of $6^o C$ from 1 January (6). Eight to 10 days after emerging the female lays eggs in clusters of five to 40 on the soil surface around the base of the plant stem. Laboratory studies have shown that most of a total of 200 eggs are laid in the first 20 days of an

average female's lifespan of 40 days (7). Eggs hatch after 7 to 10 days
and larvae burrow through the soil and enter the roots or base of the
swede bulb. Larvae mine to feed at varying depths within the bulb and
then leave the plant to pupate in the soil. Puparia are usually found in
a zone 5 cm from the skin of the bulb and 10 to 35 cm deep in the soil.
From November onwards pupae remain in diapause in the soil for 8 to 10
months before emerging as adults the following summer. TRF has diapause
requirements of several months at 4°C or below, but detailed studies have
not yet been established as for cabbage root fly (8).

Although yield losses of up to 50% fresh weight of swede crops have
been cited (1), they are usually negligible. More important is the
decrease in keeping quality of the crop in the field or in store, due to
secondary pathogens entering sites of larval mining. This in turn reduces
palatibility of the bulbs as a winter stock feed for sheep and cattle.
Recently in Scotland there has been an increase in production of table
swedes for which there is a much lower damage threshold. Chemical control
of TRF on fodder crops is not economic and on consumer crops it is often
difficult to achieve because of lack of chemical persistence in some
soils, problems of timing applications and resistance to some insecticides
(9). It is therefore desirable to breed for resistance against TRF in
swedes and turnips.

1.2 Resistance to TRF in swedes

Resistance in swedes and turnips was first noted by Morison (1) who
stated that although TRF appeared to show no preference for cultivars of
either crop, less damage was found in harder types. Recently Rygg and
Sömme (10,11) found an oviposition preference for swede compared to turnip,
and also differences in preference within cultivars of both crops. They
also found significant differences in the ratio of eggs laid to number of
larvae within the roots of different swede cultivars. Larval perception
of the host plant was found to be influenced by plant chemicals (isothiocyanates) in the skin which were identified as attractants or repellents.
Oviposition cues were also detected in swede leaf and bulb extracts,
but the effects could be stimulatory or inhibitory depending on concentration. They concluded that oviposition preference and differential
larval survival were useful components of resistance but that they
appeared to be influenced by several physical and chemical factors, some
of which were not identified.

Interest in resistance to TRF, as part of the brassica breeding programme at the Scottish Crop Research Institute (SCRI), arose after two
new swede cultivars, Angus and Melfort, were seen to have much lower
levels of TRF damage compared to other recommended types when screened by
the North and West of Scotland Agricultural Colleges (12,13,14 and Stewart,
unpublished). These two cultivars were bred at SCRI as winter hardy,
machine-harvestable replacements for Pentland Harvester, a swede susceptible to internal browning (raan). The resistance in Angus and Melfort
was later found not to be closely correlated with high dry matter as was
first suspected (15). Screening for resistance is not easy because there
are large variations in incidence and severity of attack between sites
and years. Detailed field and laboratory studies of TRF biology and
mechanisms of resistance in swedes have therefore been initiated at SCRI.
The aim is to replace laborious and inconsistent field screening of new
genotypes with a continuously available laboratory bioassay and/or a rapid
chemical screen for resistance factors. In preliminary studies in 1985
with plant breeders and chemists, we have so far concentrated on:

124

(i) field studies of oviposition periods and inter-cultivar preferences,
    (ii) egg inoculation experiments to study early stages of larval
invasion and development,
    (iii) laboratory and field assessments of damage in relation to puparia
numbers,
    (iv) mechanisms of resistance and,
    (v) development of laboratory cultures of TRF for continuous bioassay
and behavioural experiments.
Initial results at a site in N. Scotland where TRF is present at three
times the level of cabbage root fly, show that there is a marked preference for a susceptible cultivar of swede compared to Angus and Melfort
(Table I). Identification of eggs laid will show whether both or only the
dominant fly species show an oviposition preference.

Table I. Mean number of eggs laid per plant at the oviposition peak
          (28/8/85).

Cultivar	Unhatched eggs	Hatched eggs
Doon Major (S)	85.7 + 15.5	57.5 + 9.1
Angus (R)	39.2 + 6.7	25.5 + 3.7
Melfort (R)	47.6 + 12.8	26.8 + 4.4

(S) = susceptible; (R) = resistant; $\bar{x}$ + s.e., n=30.

At another site where both fly species occurred at much lower levels and
where cabbage root fly was dominant, no significant inter-cultivar oviposition preferences were found but oviposition was 10 times lower than at
the first site. Work is now in progress to study oviposition behaviour on
resistant and susceptible cultivars and to use bioassays to search for
chemical factors influencing oviposition.
    Larval invasion after egg hatch is also being assessed. During earl-
ier studies on resistance to cabbage root fly in swedes it was found that
initiation of feeding and early penetration were critical factors and that
physical and/or chemical characteristics of the skin were important components of resistance in some cultivars (16). In current studies, young
swede plants in pots are inoculated with 1 to 3 day old TRF eggs and the
development from egg hatch to larval penetration and bulb mining monitored.
Bioassays are also being used to assess the importance of the skin as a
barrier to larval invasion, although preliminary free-choice experiments
between root sections of resistant and susceptible swede cultivars with or
without skin suggest there is little discrimination at this stage. The
skin and bulb tissues of field-grown swede are being intensively sampled
during the growing season to monitor possible changes in glucosinolates
and their breakdown products when larvae first invade and after mining
within the bulb.
    Previously, field scoring of resistance in swedes to TRF has been
used to measure percentage damage and numbers of puparia of TRF and
cabbage root fly in a range of cultivars (12,13,14). Although Angus, and
to a lesser extent Melfort, show much lower levels of damage (0 to 6%)
compared with susceptible cultivars (up to 70%), differences in TRF
puparia numbers are more variable, ranging from a non-significant to a 60%
decrease on the two resistant cultivars compared to susceptible Doon Major.
Numbers of cabbage root fly puparia were not usually significantly

different between swede cultivars. It has been suggested (12,13,14) that this decrease in damage in Angus and Melfort is primarily due to shallower mining by larvae in these resistant cultivars compared to more susceptible types. More intensive sampling methods are being used to investigate the relationship between TRF and cabbage root fly numbers and damage levels in resistant and susceptible swedes.

1.3 Discussion

Preliminary results from this and other investigations suggest that the resistance of Angus and Melfort is due to the combined effects of a decrease in TRF oviposition, decreased depth of larval mining and decreased numbers of larvae developing into pupae. As suggested by other investigations, there may be several chemical and/or physical plant factors involved.

Oviposition is influenced by leaf volatile chemicals (10,11) and other surface chemicals (17,18) that may be affected by plant age, environment and plant surface microflora (19,20). Although glucosinolates and their breakdown products have been implicated in affecting the oviposition behaviour of cabbage root fly (18), other types of chemicals may be accessory oviposition activators (17). Carrot leaf waxes synergistically stimulate oviposition of carrot fly (Psila rosae F.), whilst acting as chemical defences against non-adapted bacteria, fungi, plants and herbivores (21). Physical plant characteristics may also influence oviposition. Leaf colour and total area are important in the choice of landing site for cabbage root flies after initial attraction by host plant odours (22). Visual attractiveness for oviposition may be influenced by epicuticular waxes, pubesence and leaf pigments.

Larval penetration and development inside the root may also be influenced by physical (tissue hardness, cell structure etc.) and chemical characteristics of the skin and inner tissues (10,11,16). Glucosinolates and their breakdown products may inhibit both larval perception of the host plant and development, since some of these chemicals are known to act as both insecticides and attractants (23). Other groups of plant compounds known to influence larval development of carrot fly include phenolics which have been positively correlated with larval feeding and which increase in concentration in the root after attack (24). It is important when studying plant chemistry to include the possibility of changes induced in the plant after insect challenge.

Clearly many physical and chemical plant characteristics may interact to influence levels of TRF attack. It is hoped that this research will (i) identify key stages in the life cycle of the insect and the growth stage of plant, (ii) by understanding the mechanisms involved, select one or a few key plant characters that may eventually lead to a rapid screening of genotypes for resistance and assist in genetical studies, and (iii) in the interim period, develop a laboratory bioassay to replace variable and laborious field trials relying on sporadic natural infestations at sites close to SCRI. Such bioassays may enable plant breeders to select more rapidly partial or high levels of resistance which would complement other components of integrated control of root flies. These include the use of physical barriers, crop rotations, variable planting times and plant densities, biological control by predators and parasites and possibly slow-release formulations of either secondary plant compounds or synthetic derivatives which can modify insect behaviour. Field trials in Scotland (13) have already shown that although carbofuran applied in late July adequately controls TRF under most conditions, it is necessary only on

126

susceptible swede cultivars and not economic or necessary on Angus and Melfort, even in areas of high TRF populations.

The author wishes to acknowledge the technical assistance of Mrs S.S. Lamond (SCRI) and valuable discussions with Dr S. Finch and Dr P.R. Ellis (NVRS).

REFERENCES

1. MORISON, G.D. (1937). Turnip root fly problem. Aberdeen Press and Journal, 21/1/1937.
2. SHAW, M.W. (1977). Turnip root fly control. Proc. Symp. on problems of pest and disease control in northern Britain, Dundee 1977, 62-64.
3. SHAW, M.W. (1971). Egg-laying by some dipterous pests of cultivated Cruciferae in north-east Scotland. J. appl. Ecol. 8, 353-365.
4. ANON. (1955). The West of Scotland Agricultural College Annual Report, 1955, 49.
5. JØRGENSEN, J. (1976). Biological peculiarities of Hylemya floralis Fall. in Denmark. Annales Agric. Fenniae 15, 16-23.
6. LAMB, D.J. (1984). Monitoring and forecasting the activity of root fly pests of brassicas in the west of Scotland, 1978-1983. Proc. Crop Protection in Northern Britain 1984, 228-233.
7. HAVUKKALA, I. and VIRTANEN, M. (1984). Oviposition of single females of the cabbage root flies Delia radicum and D. floralis in the laboratory. Ann. Entom. Fennici 50, 81-84.
8. COLLIER, R.H. and FINCH, S. (1983). Completion of diapause in field populations of the cabbage root fly (Delia radicum). Ent. exp. & appl. 34, 186-192.
9. TAKSDAL, G. (1966). The turnip root fly, Hylemya floralis (Fallen), resistance to chlorinated hydrocarbon insecticides in Rana, northern Norway. Acta Agric. Scand. 1966, 129-134.
10. SÖMME, L. and RYGG, T. (1972). The effect of physical and chemical stimuli on oviposition in Hylemya floralis (Fallen). Norsk ent. Tidsskr. 19, 19-24.
11. RYGG, T. and SÖMME, L. (1972). Oviposition and larval development of Hylemya floralis (Fallen) on varieties of swedes and turnips. Norsk ent. Tidsskr. 19, 81-90.
12. SHAW, M.W. (1982). Control of turnip root-fly. Research Investigations and Field Trials (RIFT) 1980-1981, School of Agriculture, University of Aberdeen, 196-197.
13. SHAW, M.W. (1984). Control of turnip root-fly. RIFT 1982-1983, School of Agriculture, University of Aberdeen, 204-206.
14. SHAW, M.W. (1985). Control of turnip root-fly. RIFT 1983-1984, School of Agriculture, University of Aberdeen, 229-231.
15. GOWERS, S., MUNRO, I.K. and GEMMELL, D.J. (1984). Turnip root-fly resistance in swedes. Cruciferae Newsletter 9, 22-23.
16. SWAILES, G.E. (1968). Feeding through root surfaces of rutabaga by Hylemya brassicae. Can. Ent. 100, 1061-1064.
17. ALBORN, H., et al. (1985). Resistance in crop species of the genus Brassica to oviposition by the turnip root fly, Hylemya floralis. Oikos 44, 61-69.
18. FINCH, S. (1978). Volatile plant chemicals and their effect on host plant finding by the cabbage root fly (Delia brassicae). Ent. exp. & appl. 24, 150-159.
19. ELLIS, P.R. et al. (1979). The influence of plant age on resistance of radish to cabbage root fly egg-laying. Ann. appl. Biol. 93, 125-131.

20. ELLIS, P.R., TAYLOR, J.D. and LITTLEJOHN, I.H. (1982). The role of microorganisms colonising radish seedlings in the oviposition behaviour of cabbage root fly, Delia radicum. Proc. 5th int. Symp. Insect-Plant Relationships, Wageningen, 1982., 131-137. Pudoc, Wageningen.

21. STADLER, E. and BUSER, H.R. (1984). Defence chemicals in leaf surface wax synergistically stimulate oviposition by a phytophagous insect. Experientia 40, 1157-1159.

22. PROKOPY, R.J., COLLIER, R.H. and FINCH, S. (1983). Visual detection of host plants by cabbage root flies. Ent. exp. & appl. 34, 85-89.

23. LICHENSTEIN, E.P., MORGAN, D.G. and MUELLER, C.H. (1964). Naturally occurring insecticides in cruciferous crops. J. Agric. Food Chem. 12, 158-161.

24. COLE, R.A. (1985). Relationship between the concentration of chlorogenic acid in carrot roots and the incidence of carrot fly larval damage. Ann. appl. Biol. 106, 211-217.

Problems encountered with the use of pheromone traps for monitoring *Mamestra brassicae* populations in Belgium

C.Pelerents
Laboratorium voor Dierkunde, Faculteit van de Landbouwwetenschappen, R.U.G., Gent, Belgium
M.Van de Veire
IWONL, Centrum voor geïntegreerde bestrijding van insekten, Faculteit van de Landbouwwetenschappen, R.U.G., Gent, Belgium

Summary

To improve efficiency of pheromone traps for monitoring the cabbage moth, *Mamestra brassicae*, its pheromone system was reconsidered. Gland extracts and collected airnborne volatiles were submitted to coupled gas chromatography-mass spectrometry.
Male olfactory and antennal responses to different isolated compounds or blends were studied by the use of different olfactometers and the electro-antennogram technique.
Prelimenary field experiments have shown the influence of trap design on its trapping efficiency.

Résumé

Problèmes liés à l'utilisation des pièges sexuels pour l'évaluation des populations de *Mamestra brassicae* en Belgique

Afin d'augmenter l'efficacité des pièges sexuels pour la noctuelle du chou il a été procédé à l'identification des phéromones par chromatographie en phase gazeuse et par spectographie de masse.
Différents olfactomètres et l'antennographie ont été employés pour étudier les réactions des mâles aux différents produits isolés par les analyses chimiques.
Des essais préliminaires en champs soulignent l'influence de la conformation des pièges sur leur efficacité.

1. Introduction

The flight period of the cabbage moth, *Mamestra brassicae* L., has been studied for many years by using light traps (VAN DAELE & PELERENTS, 1968; VAN DE STEENE & VAN PARIJS, 1985).
The first generation appears at the end of May untill begin July, and a second one at the end of July untill the end of September. The number of moths caught during the second generation was numerically much more important.
Since light traps have serious constraints another trapping method, which would reflect local variabilities in a better way, was taken into consideration.

Since 1981 we have been using delta sticky traps with formulated polyethylene caps kindly provided by Dr. Arn (Swiss Federal Research Station, CH 8820, Wädenswil, Switzerland). As the efficiency of this system was rather low in the agro-eco-system studied, we decided to reconsider the pheromone system of *M. brassicae* as well as the trapping techniques in the field.

2. Rearing of *M. brassicae*

M. brassicae is being reared in the lab since many years on an artificial diet (POITOUT, 1974). However, for this study we decided to grow the larvae on cabbage leaves. The entire rearing procedure was carried out in a room with constant temperature (25°C), a R.H. of 75-90% and a photoperiod (D/N) of 18:6.

3. Extraction, isolation and identification of *M. brassicae* pheromones

A night/day simulator apparatus was installed, enabling us to work in all seasons and during day time. By using a microprocessor steered clock, electric bulbs were commanded to lighten from N to D (and vice versa), in one hour.

To find out the pheromone production period of the females, one male was put in an activity meter during the dusk periods and night (petridish containing a capacitive bridge connection in which the slightest movement of the male caused an electric signal, which, after amplification, was recorded on paper). Males actively moved at the evening twilight, at dawn and early day time. This behaviour reflects very well the pheromone production period of the females. Based on these results, we have placed females in cages under these conditions, so we could establish by observing males that females actively produced pheromones during the dawn period. The wing position is typical and the pheromone gland is extruded. The production period of pheromones is thus rather limited.

For gaschromatographic analysis (GC), female glands were cut off at dawn using small scissors, and put into di-chloro-methane. Glands were not mixed to prevent entrance of contaminating compounds. The solution was kept 3 days at room temperature, filtrated and stored at - 20°C.

By separation on an apolar column (OV-1) 3 components were found, of which the retention times nicely agreed with those of known reference compounds (I.O.B., Instituut voor Onderzoek van Bestrijdingsmiddelen, The Netherlands) : two mono-unsaturated acetates 11-hexadecenyl acetate (11-16:Ac), 11-heptadecenyl acetate (11-17:Ac) and the saturated hexadeca-nyl acetate ($C_{16}Ac$). GC analysis using the polar column Sil88 (Chrompack) revealed that the mono-unsaturated acetates were (Z)-isomers. The (Z)-11-16:Ac, (Z)-11-17:Ac and $C_{16}Ac$ were found in a ratio of 8:1:1.

Gland extraction, however, does not necessarily give the exact reproduction of the pheromones emitted in the air. Therefore we decided to analyse air in which pheromones are "eva-porated" by females.

Females (n=15) of age 1-2 days were, prior to calling initiation, brought into a flask (1 l). An air stream of 75 ml/min was circulated continuously through a closed system during 1h30min by means of a membrane pump (Metal Bellows Corp. U.S.A.) and passing through a charcoal filter (1.5 mg) which was held in place by a filter housing (closed loop stripping).

Airborne volatiles, including pheromones are being re-tained by the charcoal. This procedure was repeated several times with other females of the same age, but using the same charcoal filter to collect sufficient amount of the pheromones. After stripping, absorbed compounds were extracted from the absorbent with 150 μl CS_2 (3x50 μl CS_2). The solution was concentrated to 5 μl, and aliquots subjected to GC and coupled GC-mass spectrometry analysis.

This led to the identification of 2 compounds: (Z)-11-hexadecenyl acetate {(Z)-11-16:Ac} and hexadecanyl acetate (C_{16}:Ac). They were found in a ratio 1:1. This ratio is quite different from those found in earlier experiments when ex-tracting pheromone glands. Herein we found (Z)-11-16:Ac and C_{16}:Ac in a 8:1 ratio, together with (Z)-11-17:Ac (VAN DE VEIRE & DIRINCK, manuscript submitted to Entomologia experi-mentalis et applicata).

4. Bioassays

Two different types of olfactometers were used. The first was a polyethylene tunnel of 4 m length with 0.6 m diameter, in which air was moved by the pulling of an exhaust fan. In the tunnel, pieces of filter paper or polyethylene (PE)-caps provided with pheromones were hung 1 m from the end. Male behaviour was checked at dawn.

Another olfactometer frequently used consisted of a wooden frame (0.5 m x 0.5 m x 0.5 m) covered with nylon gauze in which a weak air stream was created. Pheromone components or mixtures were exposed to males in the cage at dawn.

The following male responses were observed : flight to pheromone source, copulatory attempts of males with the source, copulatory attempts between males near the source and else-where.

Olfactometer studies clearly showed that young males (1-2 days) were rapidly attracted to the pheromone source. Dis-pensors (filter paper) containing 5 ng (Z)-11-16:Ac attracted male *M. brassicae*, even without using air flow. (Z)-11-17:Ac tested in several doses, did not attract males to the emittant source, but induced copulatory attempts between males, at different places in the tunnel or the cage. C_{16}:Ac and C_{14}:Ac, on the contrary, induced neither attraction nor copulatory behaviour. Male responsiveness to a formulation of 1 mg (Z)-11-16:Ac; 0.4 mg (Z)-11-17:Ac and 0.1 mg C_{16}:Ac in a polyethy-lene cap was strongly age dependent: young males (1-2 days) were attracted to the source while 5-7 day old males were not, although the latter showed copulatory behaviour.

Electroantennography (EAG) was carried out following an adaptation of the method described by ROELOFS and COMEAU (1971). Excised antennae of 1, 2, 3, 4, 5, 6 and 8 days old males were used. Two and 20 µg (Z)-11-16:Ac, were tested.

A sharp decrease in response was found when male antennae were exposed to 2 µg (Z)-11-16:Ac; antennae older than 5 days did no more respond to this dose. A slighter decrease was found when male antennae were exposed to 20 µg (Z)-11-16:Ac: antennae of 8 days old male, still elicited a distinct response. These data corresponded with those obtained by olfactometric work which revealed that young males (1-2 days) were much stronger attracted to females than older ones (5-7 days).

Antennal responses were positive to (Z)-11-17:Ac, but not to $C_{16}Ac$.

5. Field experiments

The efficiency of a sex pheromone trap is not solely dependent upon the ratio of components in the pheromone dispensor but also physical factors and trap design play an important role. Among physical parameters, temperature is very important. Extrapolation of lab results to the field is not easy. Furthermore, field results may differ from one region to another and even from one night to another.

It is also possible that high day temperatures have a negative affect on trapping , because they induce a high evaporation rate of pheromones, which in turn, could confuse male moths at dawn.

Up to 1984, delta sticky traps have been used for testing different pheromones. In 1985, funnel traps (International Pheromones Limited) were used too. The latter were more efficient for catching *M. brassicae*. This may be probably due to the presence of dichlorvos, immobilising and subsequently killing the males.

Field experiments, carried out in 1984 by Dr. Buès (INRA: Station de Zoologie et d'Apidologie, Montfavet, France) using pheromone caps from our lab, showed that water traps were much more efficient than delta sticky traps (12 x more *M. brassicae* were trapped in a 5 week period).

In 1985, during the first cauliflower growing season, three different trap designs were tested in a 0.5 ha field. The polyethylene caps were formulated with a mixture of 1 mg (Z)-11-16:Ac, 0.1 mg (Z)-11-17:Ac, 0.1 mg C_{16}:Ac and 0.05 mg (Z)-9-16:Ac.

In addition to *M. brassicae*, *Scotia puta* Hb., *Ochropleura plecta* L. and *Apamea remissa* Hb. were trapped too. Although the number of catched *M. brassicae* was rather low, a clear difference in trap efficiency was found. The water trap attracted 3.5 x more individuals than the funnel trap and 7 x more than the sticky delta trap. Bad weather conditions, especially lower night temperatures probably influenced the stickiness efficiency of the glue.

In the second cauliflower growing season preference was given to water traps, baited with (Z)-11-16:Ac (1 mg) as major component. Other caps were formulated with (Z)-11-16:Ac 1 mg,

C_{16}:Ac 1 mg (or 0.1 mg) and/or (Z)-11-17:Ac (0.1 mg). In one trap, 3 micro-capillary tubes (1 µl each) were used, instead of the usual cap dispensor. The cabbage field was square (0.5 ha), surrounded by a hedge of hornbeam (1 m height) to one side, by pasture-land to 2 other sides, and by leak and deciduous trees to the last side.

Interpretation of catches indicates an effect of surrounding vegetation; most *M. brassicae* were caught in traps along the deciduous trees (65%), while very few *M. brassicae* were found in traps close to the meadows (3%). A trap, located outside the cauliflower field, trapped not a single *M. brassicae*.

The aforementioned results do not clarify the influence of the different components on trap yields. The traps caught 8 x more *O. plecta* than cabbage moths; most *O. plecta* were found near the pasture-land in the traps, in which the smallest amounts of *M. brassicae* were observed.

The mixture (Z)-11-16:Ac (1 mg)/C_{16}:Ac (1 mg) was 5 x more attractive than the major component (Z)-11-16:Ac alone. Small numbers of *Scotia puta* and *Discestra trifolii* Hufn. were trapped too.

At the end of the experiments, residual components in the caps and microcapillary tubes were determined by GC.

The data show that release from PE-caps of (Z)-11-16:Ac is somewhat higher than those of C_{16}:Ac. The residual amount of (Z)-11-16:Ac and C_{16}:Ac after 8 weeks in the field was 40 and 50% respectively. In contrast, GC-analysis of the microcapillary tubes revealed a (Z)-11-16:Ac/C_{16}:Ac ratio of 1:5, although the initial ratio was 1:1. In the future, attention will be payed on the evaporation rate of the used components by frequent analyses.

Since rather unfavourable weather conditions in both cabbage seasons in 1985 (low temperatures, heavy rain fall) occured, interpretation of data obtained during this year may not be generalised.

6. General conclusion

By chemical analysis, pheromone components can be determined rather rapidly. Bioassays in the laboratory are suitable for detection of attractivity of one compound or mixtures of compounds.

Field experiments have shown that trap design is of utmost importance; vegetation around the experimental field and climatic condition influence also the efficiency of the traps.

Acknowledgements

Part of this research was financially supported by the IWONL (Instituut tot Aanmoediging van het Wetenschappelijk Onderzoek in Nijverheid en Landbouw).

We thank Prof. Schamp (Department of Organic Chemistry of our Faculty) for providing the analysis equipment.

Literature

1. PELERENTS, C. & VAN DAELE, E. (1968). Quatre années d'observation sur les Noctuidae de la région horticole Gantoise. Meded. Fac. Landbouwwet., R.U.Gent, XXXIII, nr. 1, pp. 1-32.
2. POITOUT, S. & BUES, R. (1974). Elevage de chenilles de vingt-huit espèces de Lépidoptères Noctuidae et de deux espèces d'Arctuidae sur milieu artificiel simple. Particularités de l'élevage selon les espèces. Ann. Zool. Ecol. anim. 6 (3), 431-441.
3. ROELOFS, W. & COMEAU, A. (1971). Sex pheromone perception: electro-antennogram responses of the redbanded leaf roller moth. J. Ins. Physiol., 17: 1969-1982.
4. VAN DE STEENE, F. & VAN PARIJS, L. (1985). Fenologische waarnemingen en chemische bestrijding van *Mamestra brassicae* L. en *Mamestra oleracea* L. in West-Vlaanderen. Landbouwtijdschrift nr. 4, Jg. 38.
5. VAN DE VEIRE, M. & DIRINCK, P. Sex pheromone components of the cabbage armyworm, *Mamestra brassicae* L.: isolation, identification and preliminary field experiments. Manuscript submitted to Entomologia experimentalis et applicata.

Monitoring of *Plutella xylostella* adults by pheromone traps – A report of a collaborative trial

J. Theunissen, J. Freuler, M. Hommes, D.J. Mowat, F. Van der Steene & F. Gfeller

Introduction

Plutella xylostella is a member of the lepidopterous pest complex in cabbage crops in western Europe. The species can reach damaging population levels in warm and dry summers but in adverse weather conditions its populations remain small. Because this potential pest does not always reach a pest status practical monitoring methods to establish its presence could be of some interest. Trapping moths using pheromone traps serves the monitoring purpose. Establishing the presence/absence of the moths is meaningful only when this has a direct relation to the infestation of the crops by caterpillars of P. xylostella. When the adults are present and anything can happen as to the rate of infestation of the crop both in severity and in time monitoring does not make sense.

To find out whether or not the results of trapping of P. xylostella adults in pheromone traps and the infestation of cabbage crops by P. xylostella caterpillars show a relationship a collaborative trial has been carried out by members of the Working Group on Integrated Control in Brassicas of the IOBC (International Organization for Biological Control).

Methods

In five countries pheromone trap catches were registered. These countries were Switzerland (Freuler and Gfeller), the German Federal Republic (Hommes), Northern Ireland (Mowat), Belgium (van der Steene) and the Netherlands (Theunissen).

During the growing seasons of 1983 and 1984 per site three traps loaded with pheromone dispensers were placed in and near cabbage fields. Two traps were placed in cabbage fields, one outside in another crop. The dispensers were loaded with a 70/30/1/10 mixture of cis 11 C16 aldehyde, cis 11 C16 acetate, cis 11 C16 alcohol and cis 9 C14 alcohol + BHT antioxydant, a synthetic pheromone of Plutella xylostella. The traps were conventional delta traps with interchangeable sticky bottoms. The dispensers were renewed once during the cropping period of the various cabbage crops. During this period the number of P. xylostella adults in the traps was registered regularly as were the numbers of P. xylostella larvae on a random sample of 10-20 plants in the cabbage crop.

Results and discussion

To determine to which extent pheromone trap catches are representative for infestation of the crop by P. xylostella caterpillars at the same location the number of days between the first catch in a trap and the first larva found in the corresponding crop have been extracted from the collected data (table 1).

Table 1: Number of days between first catch in pheromone trap and
finding of first larva in the crop.

Country	1983	1984
CH-1 (Freuler)	27	38
CH-2 (Gfeller)	53	29
D (Hommes)	35	28
NI (Mowat)	45	14
B (v.d. Steene)	7	-
NL (Theunissen)	2	15
mean	28.1	24.8
stand. dev.	20.4	10.2

Table 2: Mean catch/trap within and outside cabbage fields in two
consecutive years.

country	1983 within	outside	difference	1984 within	outside	difference
CH-1	2,5	0	2,5	40	13	27
CH-2	209	21	188	70,5	27	43,5
D	331	158	173	-	-	-
NI	29	15	14	60,5	34	26,5
B	187,5	141	46,5	-	-	-
NL	50,5	47	3,5	21	5	16

From these data it is clear that a direct relation between the moment
of catching of the adults and the first infestation of the crop is not apparent. The period between the first catch and the first recorded infestation in the field varies from 2-53 days. The mean "warning period" differs
per year and per site. This may be due to local circumstances such as exposure and weather conditions. To be used in a general early warning system
for P. xylostella infestation of cabbage crops pheromone trapping does not
give meaningful information. Since trapping results do not indicate infestation of the field in which the traps were placed they certainly are not
representative for a region other than a qualitative presence/absence indication of the adults.

While showing varying results the traps loaded with dispensers of
Plutella pheromone work well. There is a clear difference between the trapping results of traps within the cabbage crop and those outside (table 2).

The variation in trapping results is large. It may be concluded that
traps within cabbage fields catch significantly more P. xylostella adults
(P = 0,05) when the differences in both years are considered. The first
catches, which could be used as a qualitative indication of potential infestation were most often recorded from traps within a cabbage field.

Conclusion

Catches of Plutella xylostella moths in pheromone traps do show their
presence but have no temporal relationship with the infestation of local
cabbage crops by P. xylostella larvae.

Detection of movements of oligophagous Lepidoptera between host plant sowings by means of sexual traps with specific reference to *Plutella xylostella* and *Brassica* species

R.Rahn

INRA Zoology Laboratory, Le Rheu, France

Summary

Field experiments were carried out in four areas of Brittany with a view to determining the movements of diamond-back moths (Plutella xylostella) to host plants. Two varieties of Brassica oleracea - var. botrytis and acephala - and Brassica napus were studied. Sexual traps were also placed on grass and barley. The catches showed that the moths shelter under old Brassica oleracea and Brassica napus during spring from where they may fly to new Brassica sowings or young plants but not to other plants. In the case of young Brassica, they are probably attracted by the volatile flavour substances such as isothiocyanate resulting from enzymatic glucosinolate degradation.

Field experiments with genetically selected cruciferous vegetables with and without glucosinolate will be considered.

INTRODUCTION

Sexual entrapment in areas under vegetables enables the appearance of populations of certain harmful lepidoptera to be detected and their development monitored. The data obtained serve as a basis for recommendations for treatment by the Crop Protection Department and certain professional bodies (R. Rahn, 1966).

In western France, two types of vegetable are particularly well monitored, i.e. allium species, for protection against Acrolepiopsis assectella, and Brassica species for purposes of protection against Plutella xylostella (ex. maculipennis) and Evergestis forficalis.

With a view to improving forecasts, we examined several plots of the same vegetable in a particular region, together with the surrounding crops. The aim was to determine where the insects are likely to accumulate and how they infest new sowings.

EQUIPMENT AND METHODS

The traps used were the INRA type (Stockel, 1977), kept at the same height as the plants throughout the period of growth by means of an adjustable support system. The cups for the attractive cells are of the type made available to the public by the Laboratoire des Médiateurs Chimiques (chemical transmitters laboratory), Brouessy, for A. assectella and those of the Plant-Protection Research Institute, Wageningen, for P. xylostella and E. forficalis. The cells were changed by the research staff every month. In view of their cultivation patterns, sexual entrapment for Allium and Brassica crops were carried out as follows:

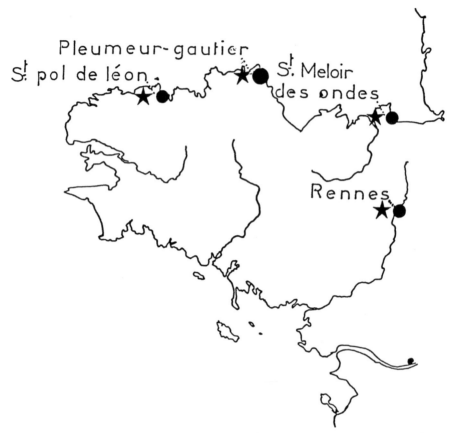

Fig. 1. Distribution of Sexual Traps for Two Species of lepidoptera which
attack <u>Brassica</u> species in the vegetable growing area of Brittany -
<u>Plutella</u> <u>xylostella</u> and <u>Evergestis</u> <u>forficalis</u>

- in second year plots the host plants;
- in first year plots destined for vegetable production;
- in the crops surrounding the host plant plots.

The network of traps comprised four stations spread over the coastal
vegetable production region of Brittany - at St Pol de Léon (Finistère),
Pleumer Gautier (Côtes du Nord), St Mélior des Ondes (Ille et Villaine) and
on the outskirts of Rennes in the INRA area of the Plant Improvement Station
(cauliflower and fodder kale laboratories). Cf. Figure 1.

The traps were checked three to five times per week. In the case of
accidental capture of large numbers of unwanted insects (diptera attracted
by the organic products developing, for example), the bottoms of the traps
were changed or recoated with bird lime after cleaning. At Rennes, two
phenological characteristics of the cabbage were noted, i.e. the number of

---

* The use of birdlime available in an aerosol container under the name
'Soveurode' from the chemical company Sovilo very much simplifies this
operation.

Catches

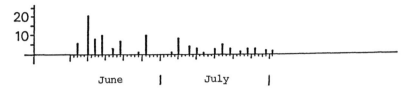

Fig. 2. Catches of P. xylostella with Sexual Traps Set in Old Cabbage
Plantings. St. Pol de Léon 1985.

P. xylostella

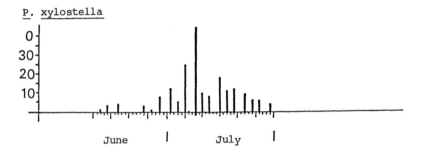

Fig. 3. Male P. xylostella Caught by Means of Sexual Traps Placed in
Cauliflower Nurseries - St Pol de Léon 1985.

leaves and the height of the plant. This document describes the results
obtained with Plutella xylostella.

RESULTS

1. Sexual Entrapment at Saint Pol de Léon

1.1   Second year cauliflower plots

The trapping, which was carried out during June and July before
harvesting of the plants, showed that P. xylostella were constantly present.
The greatest number of insects were caught in the first half of June, and
the figures dropped off steadily in the course of July (cf. Figure 2).

1.2   Cauliflower Nursery

The catches made at the same time in a cauliflower nursery show that,
in this same geographical sector, P. xylostella are much more active in July
(cf. Figure 3).

2. Results at Pleumeur-Gautier

Two traps were used simultaneously - one in a cauliflower nursery and
the other on a nearby grassed area. 151 males were captured among the
cauliflowers as against a mere ten on the grass (cf. Figure 4).

139

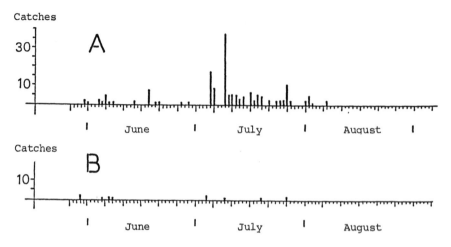

Fig. 4. Effects of Vegetation on Behaviour of P. xylostella: Catches on Host
       Plants (Young Cabbages - A) and Non-Host Plants (Grass - B).
       Pleumeur-Gautier 1985.

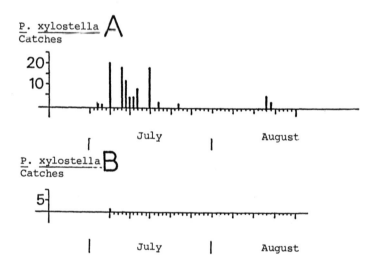

Fig. 5. Effects of Vegetation on Behaviour of P. xylostella: Catches on Host
       Plants (Young Cabbages - A) and Non-Host Plants (Barley - B). St
       Méloir des Ondes 1985.

3. Results at St Méloir des Ondes

       At the La Rimbaudais farm the two traps were placed 30 m apart among
cauliflowers, planted that same year, and barley on one and the same lot of
open land. Cultivation of various species of Brassica is widespread in this
sector. In the cauliflower nursery, P. xylostella were found to be very
active in July, 95 males being caught as against only one among the barley
(cf. Figure 5).

Catches

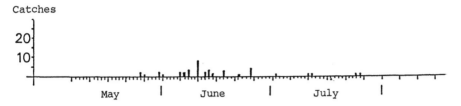

Fig. 6. P. xylostella Caught With the Aid of Sexual Traps on Colza (Brassica napus). Le Rheu 1985.

Catches

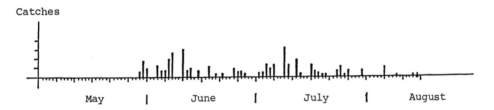

Fig. 7. P. xylostella Caught With the Aid of Sexual Traps in Fodder Kale Nurseries (Brassica oleracea var. acephala). Le Rheu 1985.

4. Results at Le Rheu

At the INRA experimental farm we set traps for P. xylostella among three varieties of Brassica: a cauliflower known as B. oleracea var. botrytis, and two species of Brassica used as animal feeding stuffs, i.e. B. oleracea var. acephala (fodder kale) and B. napus (colza).

4.1 Catches of P. xylostella on Brassica napus

The trap, which was placed at the beginning of May, demonstrated the flight of P. xylostella which takes place during June, in spite of relatively limited catches, i.e. a total of 38 males between 8 May and 2 August, five of which were caught in July and none in August (cf. Figure 6).

4.2 Captures of P. xylostella

The different ways in which B. oleracea var. acephala are grown made it possible to maintain traps set on different dates.

a) Sexual traps in nurseries

No insects were caught in the traps set on 7 May until May 28. From then on catches continued until beginning of August, which reflects two flights, the first in the first half of June and the second a month later. A total of 175 males were captured (cf. Figure 7).

b) Traps in later sowings

The trap was maintained from 18 June to 14 August. The July flight was very clearly detected in that 317 males were captured (cf. Figure 8).

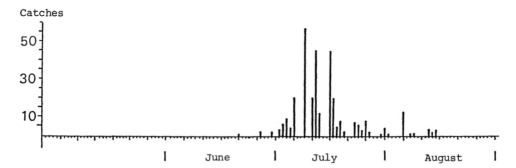

Fig. 8. P. xylostella Caught With the Aid of a Sexual Trap Placed on a
Fodder Kale (B. oleracea var. acephala) Seed Bed. Le Rheu 1985.

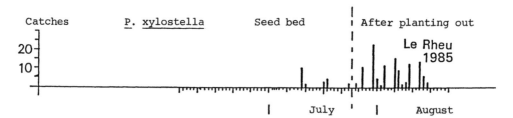

Fig. 9. P. xylostella Caught With the Aid of a Sexual Trap Placed on the
Seed Bed and Subsequently in the Field After Planting Out. Le Rheu
1985.

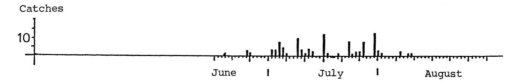

Fig. 10. P. xylostella Caught With the Aid of a Sexual Trap Placed on Second
Year Summer Cauliflower (B. oleracea var. botrytis). Le Rheu 1985.

c) Traps on plantings under forcing frames and after planting out

The position of this lot, sheltered between the laboratories and the
greenhouses, and the artificial conditions resulting from the use of forcing
frames meant that very few insects were captured, i.e. 19 over 28 days from
the beginning of June. However, immediately after planting out, adult
insects were attracted, 120 males being captured over the next 22 days (cf.
Figure 9).

4.3 Captures of P. xylostella on cauliflowers of the species Brassica
     oleracea var. botrytis

Two traps were set. With the first, among summer cauliflower, regular
but small catches were made, i.e. 91 males between 18 June and 14 August
(cf. Figure 10).

142

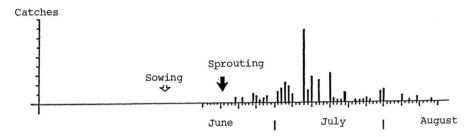

Fig. 11. P. xylostella Caught With the Aid of a Sexual Trap Placed on a
Cauliflower Seed Bed (B. oleracea var. botrytis). Le Rheu 1985.

With the second trap, set in a nursery, the first catches were made
immediately after sprouting, the July flight being clearly demonstrated with
the capture of 128 males (cf. Figure 11).

DISCUSSION AND CONCLUSIONS

As soon as the insects resume activity in spring, old sowings of host
plants can shelter adults which have either hibernated there or have
returned after hibernating nearby. These plants from the previous year - St
Pol de Léon 1985 and the species of Brassica intended for animal feedstuffs,
such as Brassica napus, Le Rheu 1985 - allow P. xylostella to develop to the
adult stage. After this point the host plants soon flower and go to seed.
The vegetative tissues become leathery and even the young siliques, which
are tender when they originally form, dry out quickly. At this stage,
Brassica species are unsuitable hosts for a new generation of the pests.
Thus, the populations of P. xylostella move to plantings which provide
growing vegetable tissues for a sufficient time - e.g. St Pol de Léon 1985,
Figures 2 and 3. These tissues may be very young. In all the cases in which
the traps were set at the time of sowing, the first catches of P. xylostella
were made as soon as the first cotyledons had appeared - indeed even when
they were emerging from the ground (e.g. St Pol de Léon No.3 Le Rheu No.11).
Of the various factors affecting the choice of new plantings, visual
signs are known to play a part with homoptera (V. Moericke, 1979) and
diptera (Brunel et al., 1970, 1975 and 1981). Visual signals affect the
behaviour of the lepidoptera, such as the pieridae. However, the movements
or reproduction patterns of insects of this order are frequently triggered
by the detection of volatile products released by the plant - e.g. Allium -
and the sexual partner (pheromones) (R. Rahn, 1968), and this is probably
the area to be investigated with a view to explaining the link between host
plant and pest in the case of Brassica, since it would appear unlikely that
visual signals alone could attract the insects while the tiny leaves are
still partly buried. The cruciferae, however, are rich in glucosinolates,
which, in the presence of myrosinase, produce isothiocyanates, thiocyanates
and nitriles by enzymatic degradation. Isothiocyanates - which are also
found in Allium species - are highly volatile products which affect the
behaviour of adult insects (Auger and Thibout, 1979; Le Comte and Thibout,
1981).
Monitoring the phenological stages of the plants showed that the times
when the largest number of adults can be captured correspond to those when
young leaves are forming. An analysis of volatile products in cabbage leaves
as a function of the position of the leaves shows a concentration of
allylisothiocyanate of 16.5% in the inner leaves as against 3% elsewhere

143

(Macleod, 1976). Thus the youngest plants are likely to produce the largest quantities of attractive products. Since these plants provide the ideal diet for young larvae, adult P. xylostella and other pests, such as Evergestis forficalis have, over the course of evolution, adopted this chemical signal as a means of locating the host plant. Since this same signal promotes the secretion of the sexual pheromone by the female, reproductive behaviour, from mating to laying, is combined with ideal survival conditions.

Since the volatile products result from the decomposition of glucosinolates, we plan, in collaboration with geneticists working on cabbage and cauliflower, to compare the attractivity of strains of Brassica which are rich or poor in glucosinolates. If these levels are found to affect the relationship between host plants and pests, it might be possible to produce strains which are not attractive, or at least less so, to their specific enemies.

REFERENCES

(1)   AUGER, J. and THIBOUT, E., 1979. Action des substances soufrées volatiles du poireau (Allium porrum) sur la ponte d'Acrolepiopsis assectella (Lepidoptera: Hyponomeutoïdea) prépondérance des thiosulfinates. Can. J. Zool. 57, 223-229.

(2)   BRUNEL, E. and LANGOUET, L., 1970. Influence de caractéristiques optiques du milieu sur les adultes de Psila rosae Fab. (Diptères Psilidés): attractivité de surfaces colorées, rythme journalier d'activité. C.R. Soc. Biol. 164, 7, 1638-1644.

(3)   BRUNEL, E. and BLOT, Y., 1975. Rôle de la couverture végétale sur les captures de Psila rosae Fabr. (Diptères Psilidés) au moyen du piège jaune. Sci. Agro. Rennes, 91-96.

(4)   BRUNEL, E., FREULER, J. and DUCHESNE, J., 1981. Determination des variations de la réflectance du feuillage de quelques ombellifères en fonction de l'âge et de la variété: conséquences sur le comportement de Psila rosae. Coll. signatures spectrales d'objets en télédetection, Avignon 8-11 September.

(5)   LE COMTE, C. and THIBOUT, E., 1981. Attraction d'Acrolepiopsis assectella en olfactomètre par des substances allélochimiques volatiles d'Allium porrum. Ent. Exp. et Appl., 30, 293-300.

(6)   MACLEOD, A.J., 1976. Volatile flavour compounds of the crucifere p. 307-328 in the Biology and Chemistry of the Cruciferae. Ed. VAUGHAN, MACLEOD and JONES (1976) Academic Press London.

(7)   MOERICKE, V., 1969. Host plant specific colour behaviour by Hyalopterus pruni (Aphididae). Ent. exp. Appl., 12, 524-534.

(8)   RAHN, R., 1966. La teigne du poireau Acrolepia assectella Zeller, éléments de biologie et mise au point d'avertissements agricoles fondés sur le piégeage sexuel des mâles. C.R. Acad. Agric. Fr., 997-1001.

(9)   RAHN, R., 1968. Rôle de la plante-hôte sur l'attractivité sexuelle chez Acrolepia assectella Zeller (Lep. Plutellidae). C.R. Acad. Sc. Paris 266, 2004-2006 (6 May 1966).

(10)   STOCKEL, J., 1977. Modifications apportées au piège sexuel INRA pour les insectes. C.R. réunions 'Les phéromones sexuelles des insectes'. Montfavet, 25-27 October.

Studies of soil receptiveness to clubroot caused by *Plasmodiophora brassicae*: Experiments on responses of a series of vegetable soils in Brittany

F.Rouxel & M.Briard
INRA, Station de Phytopathologie, Domaine de la Motte-au-Vicomte, Le Rheu, France
B.Lejeune
INRA, Laboratoire d'Amélioration des Plantes Légumières Plougoulm, Saint Pol de Léon, France

Summary

Although clubroot (caused by Plasmodiophora brassicae) is one of the most damaging diseases afflicting cauliflower crops, not much is known about its actual extent and the interaction between parasite and soil factors. This paper starts by giving the findings of a study to estimate soil infectivity and discover the extent and distribution of the disease on a 100 hectare vegetable growing area divided into 160 parcels. According to the study, 40% of the parcels are infected with the disease to varying degrees. By comparing the clubroot receptiveness of 14 different soils, it was shown that future infection risks for uncontaminated soils differ greatly. In general terms, soils are at greater risk the more acidic they are, but pH levels alone do not explain every case. It seems, in particular, that autoclaving increases receptiveness to different extents, depending on soil type. The cumulative effect of the two associated treatments (acidification + autoclaving) indicates that a specific type of microflora might influence the parasite's life cycle, thus encouraging development of the disease in an acidic environment.

1. INTRODUCTION

Clubroot, caused by the soil fungus Plasmodiophora brassicae, is one of the major parasites infecting cauliflower crops in Brittany and many other vegetable growing regions of France (1). This disease, which has been around for a long time, can show up at the nursery stage or in the field after transplanting, but it is usually the early stage attacks which do the most damage. At the moment there is no known pest management method which is both effective and easy to apply: all the different varieties and hybrids of commercial cauliflowers are susceptible, and chemical control methods are very much a hit-and-miss affair despite the large amount of research carried out throughout the world. Therefore, the only practical solution is to be sought in the growing conditions, which are known to influence development of the disease, i.e. soil acidity and moisture content, frequent cultivation of cruciferae on the same parcels (2, 3, 4). Experience shows, however, that it is difficult to get to grips with all these interacting factors in practice, which explains why farmers are often unable to predict the appearance of the disease.

For this reason we deemed it worthwhile to study, at micro-region level, the actual extent and distribution of the parasite by estimating soil infectivity. Secondly, we wished to find out whether soils currently free from clubroot have a predisposition to non-contamination, and if so, why. To do this we compared the receptiveness levels of various soils to the disease. This work, aimed at finding a way of predicting soil-linked risks to some extent, dovetails with research to find an overall control method.

2. MATERIALS AND METHODS

The study took place at Plouenan, Brittany, in the heart of the vegetable growing region of North Finistère. It covers 160 parcels of land (totalling some 100 hectares) forming a comparatively homogeneous zone pedologically and agronomically speaking: deep sandy loam soils, and crop rotation cycles characteristic of the region - principally cauliflower/artichoke.

2.1   Estimating Soil Infectivity

The extent of P. brassicae in the soil was estimated by trapping, using a plant very sensitive to the parasite. About 15 litres of soil were taken from 20 sample plots along the two diagonals of each of the 160 parcels. After homogenisation and measurement of pH (water), the sample was divided up into two seed pans which were seeded (50 seeds per pan) with Chinese cabbage (Brassica campestris, Granaat variety). The crop was grown under glass for six weeks in conditions favourable to development of the disease (t° = 18-25°C, very moist soil), after which the plants were taken out and divided into four classes according to the degree of infestation. The four classes were assigned coefficients (0, 0.25, 0.5 or 1). It was then possible to calculate the 'Disease Index', which reflects the average infectivity of the soil on each parcel.

2.2   Comparing Soil Receptiveness to Clubroot

Comparisons were made (5) of the appearance of the disease on a number of sensitive plants following artificial soil infestation under controlled conditions. After drying in the ambient air, each sample was crushed and homogenised and then inoculated with increasing doses of P. brassicae. The growing conditions of the Chinese cabbage and the methods of recording results were identical to those used in estimating soil infectivity. The Disease Index values, expressed as a function of the various inoculum doses, reflect the level of receptiveness of each soil.

The behaviour of 12 soils with zero infectivity was thus compared to that of a control known to be very receptive and to that of a soil with high infectivity (DI = 45%). The P. brassicae isolate used came from root galls of cauliflowers grown at Plouenan (pathotype No. 16-31-28, determined in accordance with the ECD test of Buczacki et al. (6)). The final inoculum concentrations were 10^4, 10^5, 10^6, 10^7 and 10^8 resting spores per ml of soil. The test was carried out using five 300 ml pots for each dose level.

2.3   Study of the Influence of Some Soil Factors on Receptiveness

Two factors were investigated, on their own or in association:
the pH: a not very receptive soil, with an initial pH of 7.4, was acidified by adding sulphuric acid diluted to 1% in order to give pH levels of 6.5, 5.5 and 4.5;

steam _treatment of the soil_: three soils, with zero infectivity and fairly low levels of receptiveness, were autoclaved (110°C, 30 minutes) twice, with an interval of 48 hours in between.

One week later the effect of these treatments on clubroot receptiveness was measured using the method previously described.

3. RESULTS

3.1 Extent and Distribution of Clubroot in the Area Studied

The infectivity estimates (Table 1) show that 40% of the parcels are affected by the disease, which is a high proportion when compared to traditional estimates for crops. However, the levels of attack vary considerably, with only 11% of the parcels having infectivity levels over 20%.

TABLE 1 Importance and Distribution of Clubroot in the 160 Analysed Parcels, in Relation with Soil pH Levels

Disease Index [1] classes	Parcels		Soil pH [2]
	number	%	
0	96	60	7.2 ± 0.95
1-20	46	29	6.6 ± 0.64
> 20	18	11	6.5 ± 0.63

[1] Measured in controlled conditions
[2] Average soil pH ± standard deviation.

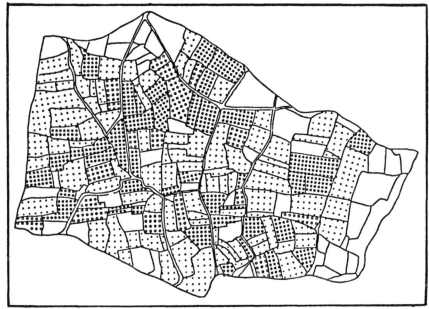

Fig. 1. Geographic Distribution of Clubroot in a 100 hectares Area in Brittany
Parcels with Disease Index = 0 [∴∴∴] 1-20% [⋮⋮⋮⋮] > 20% [⋮⋮⋮⋮]

147

The parcels affected are spread over the whole area studied (Figure 1), and although there seems to be a certain Y-shaped concentration within the area, this distribution pattern does not coincide with any particular topographical or agronomic feature. The relation between development of the disease and soil acidity is confirmed, however, because the average pH of the affected parcels is around 6.5 whereas that for unaffected parcels is 7.2. When the disease is present the degree of infestation does not seem to be related to pH values, but more to repeated cultivation of cruciferae, which increases the amount of parasite inoculum in the soil.

3.2 Comparison of Soil Receptiveness to Clubroot

Given the preceding results, the question arises as to why no clubroot is to be found in 60% of the parcels located in the heartland of a traditional vegetable growing area: is it because there is no parasite contamination, or because certain soils are less receptive to the disease? The pH factor alone seems unable to account for all cases, as shown by the wide variations in pH levels in both infestation and non-infestation conditions (Table 1: standard deviations between 0.6 and 0.9).

TABLE 2 Comparison of 14 Soils for Receptiveness to Clubroot: Distribution in Three Classes After Measuring Disease Index on Chinese Cabbages Cultivated in Soils Infested by Increasing Inoculum Levels of P. brassicae

Parcel	Soil pH	Resting spores of P. brassicae/ml of soil						
		0	10^4	10^5	10^6	10^7	10^8	
MEVEL	8	0	0	6	11	70	62	CAT. III
LEBER	8.3	0	0	1	21	55	72	
CAZUC	8.2	0	0	6	51	40	76	
MESMEUR	6.1	0	2	7	39	70	84	
ROUE	8	0	0	12	54	86	92	CAT. II
MADEC	6.7	0	0	12	39	95	93	
PAUGAM	7.5	0	10	32	66	97	89	
PRIGENT	7.2	1	6	47	62	93	93	
CREACH	5.9	0	30	98	92	76	100	
PENNORS	6.1	1	60	92	95	95	100	CAT. I
CREIGNOU	7.4	0	75	87	92	94	100	
MERCIER	5.2	0	89	90	88	87	100	
CONTROL	5.1	5	98	98	96	100	100	
GUIVARCH	5.2	45	91	100	100	100	100	

Table 2 gives the results obtained in comparing the receptiveness of 14 soils: 12 selected from the 96 with zero infectivity and two controls. The 13 soils naturally free of the disease can be divided into three categories according to the inoculum dose required to produce a Disease Index value >10%. The five soils in category I are very receptive, since a weak dose of inoculum (10^4 spores per ml of soil) is enough to cause high, and sometimes very high (DI almost 100%), infestation. At the other end of the scale, category III contains four much less receptive soils, since a dose of inoculum 100 times more concentrated is needed to cause infestation levels of any significance, while an inoculum dose 10 000 times more concentrated (10^8 spores per ml) is needed to cause comparable levels of attack. The behaviour of the four soils in category II falls between that of the other two.

Comparison of these results with pH values is very revealing, showing as it does an inverse relationship between pH values and receptiveness levels similar to the observations already made for infectivity. There are, however, a number of exceptions, with an acid soil (Mesmeur, pH = 6.1) showing low receptiveness while an alkaline soil (Creignou, pH = 7.4) is very receptive.

3.3   Relations Between Some Soil Factors and Receptiveness to Clubroot

The above results first of all prompted us to try to verify experimentally on one type of soil whether there is a link between pH and receptiveness to clubroot. Figure 2 clearly shows that increasing the acidity of a soil initially possessing a high pH (changing from 7.4 to 4.5) increases its level of receptiveness.

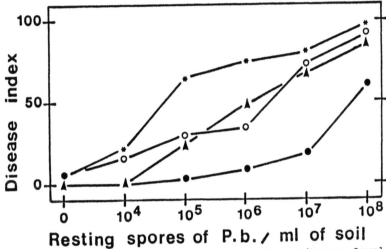

Fig. 2. Relationships Between Soil Acidity and Receptiveness Level to Clubroot: Soil at Initial pH = 7.4 Acidified Progressively by Diluted Sulphuric Acid

pH 7.4 ●   6.5 ▲
    5.5 ○   4.5 *

Subsequently, we tried to find out whether the biological component of the soil might play a role and explain the cases where receptiveness levels do not correlate to pH levels. Figure 3 shows that autoclaving three soils increases their receptiveness to clubroot, but to varying degrees. Although the heat treatment was not accompanied by microbiological analyses to monitor microflora development, it is feasible that the increase in receptiveness of soil 1 (which was greater than that of soil 3) is due more to biological factors, whereas the increase in soil 3 is due more to physico-chemical factors. These differences in behaviour following autoclaving might be related to acidity, since the pH of the three soils differed considerably.

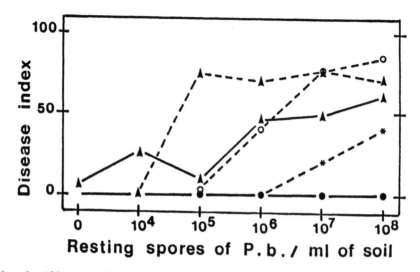

Fig. 3. Effect of Steam Treatment (110°C, 30 minutes) on Receptiveness
Levels of Three Soils:

    soil 1 : pH = 7.7 o
    soil 2 : pH = 6.1 ▲
    soil 3 : pH = 8.2 *

    non steamed soil ———   steamed soil — — —

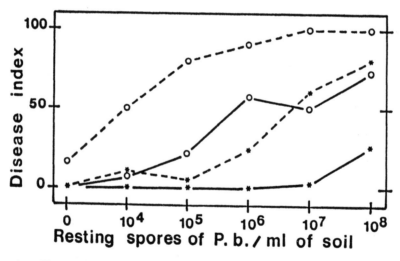

Fig. 4. Effect of Steam and Acidification Treatments (Alone or Associated)
on Receptiveness Level of a Soil to Clubroot

    pH 8.2 - non steamed soil   ——— * ———
    pH 4.5 - non steamed soil   ——— o ———
    pH 8.2 - steamed soil   — — * — —
    pH 4.5 - steamed soil   — — o — —

150

We carried out a third experiment in which we studied the combined impact of both factors. The results of Figure 4 confirm the influence of acidity and autoclaving on receptiveness, the effects of these two factors being comparable under our experimental conditions. However, combination of the two treatments clearly raises soil receptiveness levels, which leads us to suspect a link between soil pH and microflora.

4. DISCUSSION AND CONCLUSIONS

The results set out in this paper detail the extent and distribution of clubroot in a vegetable growing area of 100 hectares. The proportion of contaminated parcels is larger than is generally held and does not necessarily reflect the actual incidence of the disease among the crops, given the important effect of climatic factors (7) on parasite development. Nevertheless, it has to be admitted that the P. brassicae inoculum is actually present in all the parcels with positive infectivity, and there is a risk of it increasing under the impact of various factors (short rotation periods, insufficient drying of the soil due to certain crop management practices).

The results also show that the risk of developing the disease varies for the as yet uncontaminated soils. Thanks to the receptiveness comparisons it was possible to divide the soils into three categories based on their predisposition to clubroot, thus making it possible to develop a method of predicting soil-linked risks for the medium term. It would appear worthwhile extending this type of study to other crop growing areas more diversified in both pedological and agronomic terms.

For a long time now the appearance of clubroot has been associated with certain soil factors such as pH and moisture content, but without the mechanisms involved being fully appreciated. This explains why the relevant literature still contains contradictions despite recent work in this field (8, 9, 10, 11). In overall terms, our results confirm the link between the presence of clubroot and soil acidity, but there seems to be no correlation between infectivity levels and pH values. By contrast, there is an inverse relationship between pH values and soil receptiveness levels, as we have already shown for colza (5). Therefore, one can conclude from this that soils are more 'sensitive' to the disease the more acidic they are, but once the 'sensitivity threshold' has been crossed it is the amount of the parasite inoculum, due to frequent cultivation of cruciferae, which determines the soil infectivity level and thus the immediate risk of disease.

Moreover, the receptiveness comparisons helped to confirm observations made in the field, i.e. that clubroot develops in certain soils with high pH levels. Study of such soils under controlled conditions (using the same P. brassicae isolate for all the soils compared) leads us to reject the hypothesis that certain parasite strains adapt themselves to soils with high pH levels. It seems that other factors in such soils affect the development of the disease.

It has been shown that autoclaving increases receptiveness to different degrees according to the type of soil. Decreasing pH level and then autoclaving has a cumulative effect which seems to point to some interaction between physico-chemical and biological factors. In the absence of microbiological analyses it would be unwise to speculate on the mechanisms involved here. The main aim must be to try to confirm the results obtained and to look for correlations between changes in receptiveness and the quantitative and qualitative development of the soil microflora following various kinds of treatment.

Recent works (12, 13, 14) have analysed the composition of the walls of resting P. brassicae spores. In view of the high protein levels discovered it is feasible that proteolytic microflora in the soil help to break down these walls, thus contributing in part to the appearance of the fungus. If this is shown to be the case, studies of this type of microflora should lead to a better understanding of how the P. brassicae inoculum develops in soils, and should undoubtedly help explain why receptiveness is high in some cases but not in others.

REFERENCES

(1)   JOUAN, B.   and   ROUXEL, F., 1979.   Les   principales   maladies   du Chou-Fleur. 139-164 - in 'Le Chou-Fleur' - INVUFLEC, Paris.

(2)   KARLING, J.S., 1968.   The Plasmodiophorales,   256 p. HAFNER Publishing Company, London.

(3)   ROUXEL, F. and JOUAN, B., 1982.   La Hernie des Crucifères. 74-84 - in 'Les maladies des Plantes'. ACTA, Paris.

(4)   BUCZACKI, S.T., 1983.   Plasmodiophora.   An inter-relationship between biological and practical problems.   161-191 - in Zoosporic Plant Pathogens. A modern perspective. BUCZACKI Ed. Academic Press, London.

(5)   ROUXEL, F. and REGNAULT, Y., 1985.   Comparaison de la réceptivité des sols à la Hernie des Crucifères: Application à l'évaluation des risques sur quelques sols à culture de Colza oléagineux. 375-382 - in C.R. des Premières Journées d'études sur les maladies des plantes. A.N.P.P., Versailles.

(6)   BUCZACKI, S.T.,   TOXOPEUS, H.,   MATTUSCH, P.,   DIXON, G.R.,   and HOBOLTH, L.A., 1975.   Study of physiologic specialization in Plasmodiophora brassicae; proposals for attempted rationalization through an international approach. Trans. Br. Mycol. Soc., 65, 295-303.

(7)   THUMA, B.A., ROWE, R.C., and MADDEN, L.V., 1983. Relationships of soil temperature and moisture to Clubroot severity on radish in organic soil. Plant Disease, 67, 7, 758-762.

(8)   HAMILTON, H.A. and CRETE, R., 1978.   Influence of soil moisture, soil pH, and liming sources on the incidence of Clubroot, the germination and growth of cabbage produced in mineral and organic soils under controlled conditions. Can. J. Plant Sci. 58, 45-53.

(9)   FLETCHER, J.T.,   HIMS, M.J., ARCHER, F.C. and BROWN, A., 1982. Effects of adding calcium and sodium salts to field soils on the incidence of Clubroot. Ann. Appl. Biol. 100, 245-251.

(10)   CAMPBELL, R.N.,   GREATHEAD, A.S., MYERS, D.F. and DE BOER, G.J., 1985. Factors related to control of Clubroot of Crucifers in the Salinas Valley of California. Phytopathology, 75, 665-670.

(11)   MYERS, D.F. and CAMPBELL, R.N., 1985. Lime and the control of Clubroot of Crucifers: Effects of pH,   Calcium,   Magnesium,   and their interactions. Phytopathology, 75, 670-673.

(12)   MOXHAM, S.E. and BUCZACKI, S.T., 1983. Chemical composition of the resting spore wall of Plasmodiophora brassicae. Trans. B. mycol. Soc., 80, 297-304.

(13)   MOXHAM, S.E.,   FRASER, R.S.S.   and BUCZACKI, S.T., 1983.   Spore wall proteins of Plasmodiophora brassicae. Trans. Br. mycol. Soc., 80, 497-506.

(14)   BUCZACKI, S.T. and MOXHAM, S.E., 1983. Structure of the resting spore wall of Plasmodiophora brassicae revealed by electron microscopy and chemical digestion. Trans. Br. mycol. Soc., 81, 221-231.

Session 2
Carrot crops

Chairman: R.Cavalloro

Field and laboratory studies on the behaviour of the carrot fly, *Psila rosae*

G.Skinner & S.Finch
National Vegetable Research Station, Wellesbourne, Warwick, UK

Summary

Spring ploughing did not reduce the numbers of carrot flies that emerged from fields infested with this pest. Water-traps were effective for monitoring fly activity in the field, though the most appropriate hue of yellow paint needs to be confirmed. The time of activity of the first generation was forecast accurately using the number of day-degrees accumulated above a base temperature of 6°C. Further work is required to produce a correspondingly accurate forecast for the second generation. Standard sampling methods are essential, if data collected by Group members are to be incorporated into any generalized model for forecasting carrot fly attacks.

1. INTRODUCTION

During the last five years a concentrated effort has been made to produce a system for accurately forecasting the field behaviour of the cabbage root fly (1). The aim of the present research is to produce a comparable system for the carrot fly. Hopefully, the present research will gain by avoiding many of the pitfalls of the earlier study.

Although the carrot fly has been studied at Wellesbourne since 1950, the work has been restricted mainly to evaluating insecticidal (e.g. 2 & 3) and plant-resistance methods of control (4). Biological data has been collected only in years when time permitted. The major reason for the paucity of detailed biological data has been the inability to maintain a robust <u>Psila rosae</u> colony under laboratory conditions.

The current work therefore has been restricted either to field studies or to laboratory studies using field-collected insects. The work has concentrated on 1) assessing the rate of build up of carrot fly populations under natural conditions, 2) the factors regulating mortality in the field, 3) the development of traps for monitoring fly numbers and, 4) the use of such traps to determine the peak activity of flies in the field. This paper is really a progress report and is intended as a guide to the type of work in which we could carry out co-operative experiments.

2. EXPERIMENTAL WORK

2.1. Build up of carrot fly populations

Crop isolation is one way of avoiding high infestations of the carrot fly. The present research was started to see how long it would take the carrot fly to build up into damaging numbers in an area previously free of carrot fly.

A 0.5 ha plot of Vita·Longa, a cultivar grown commonly at
Wellesbourne, was drilled on 11 April 1981 in a field approximately 1 km
away from fields used routinely for experiments on this pest. Inspection
of roots during the summer indicated that carrot flies had entered the
crop but few were caught in yellow traps. In December 1981, pupal
samples were taken from 20 cm of row. Samples were taken every 6 m apart
across the beds and 10 m apart along the beds. Approximately half of the
66 samples contained pupae. Edge effects were most pronounced along the
eastern edge of the crop, where it bordered onto a crop of mowing-grass,
Lolium perenne (L.). Along this edge of the crop there was a mean number
of 51 pupae/m^2, with a maximum of 133 pupae/m^2 occurring in one corner.
Throughout the main part of the crop there were localized areas with up
to 27 pupae/m^2. Overall, the carrot fly population averaged 17 pupae/m^2
or the equivalent of 170,000 pupae/ha. Interpolating from the nomogram
of Wheatley and Freeman (5), such an infestation would damage about 5% of
the crop.

To continue the study in 1982, a similar plot of Vita Longa was
drilled adjacent to the 1981 plot on 16 April 1982. In December 1982,
half of the 66 soil samples again contained carrot fly pupae. Crop
damage was heaviest along the western edge of the plot where it bordered
a crop of swedes. Along this edge there were up to 60 pupae/m^2 and 42%
of the carrots were mined. Overall there were similar numbers of pupae
($18/m^2$) to those recorded in 1981 ($17/m^2$). Hence, although the summer
did not favour the build-up of carrot fly infestations, once the pest was
established it was capable of maintaining its previous population level
even under adverse conditions.

2.2. Factors effecting mortality in the soil

Some carrot fly pupae can be buried up to 30 cm deep by ploughing.
To determine the effect of depth of pupae on subsequent insect survival,
batches of healthy and parasitized pupae were buried in 7.5 cm pots 5,
10, 20, 25 and 30 cm deep. The pupae were buried in Tygan sleeves made
from mesh sufficiently fine to exclude beetle predators (6). Although
45% of the carrot flies and 55% of the parasite Chorebus gracilis (Nees)
emerged from depths of 5 and 10 cm, only 12% of both insects emerged from
depths of 30 cm. Inspection of the pupae revealed that similar numbers
ecclosed at all depths and that mortality occurred as the flies tried to
burrow out of the soil.

To determine the effects of ploughing on pupal mortality, half of a
field plot containing a high density of carrot fly pupae was ploughed on
30 March 1981. A cage 3.1 m x 6.1 m x 2 m high was then erected on the
ploughed area and a second cage was placed over the adjacent area of
unploughed land. The 2 mm aperture mesh of these cages retained the
carrot flies, but was not sufficiently fine to retain the parasites. To
obtain a better estimate of the numbers of parasites that emerged, six 30
cm x 30 cm x 30 cm cages made from 1 mm aperture mesh were erected over
groups of surviving carrots inside the larger cages. The total numbers
of flies/parasites caught from 1 May to 20 July in the larger cages over
the ploughed and unploughed areas were 4952/567 and 4548/496,
respectively. The similarity in the numbers of insects recovered
indicated that spring ploughing has no effect in reducing carrot fly
numbers. Totals of 1647 carrot flies were caught/m^2 from the 1.2 m^2 of
cage covered by the twelve small cages, compared to only 257/m^2 from the
remaining 37 m^2 of the two·larger cages. This provided a clear

demonstration of how easy it is to overestimate the numbers of carrot flies in a field when samples are not taken at random. By assuming that similar proportions of carrot fly were parasitized in the large as in the small cages, it was estimated that 330 carrot flies and 140 parasites emerged from each m^2 of soil, the equivalent of 2-3 larvae/carrot (5).

2.3. Development of traps for adults

Experiments to determine the most attractive colour for carrot fly traps were carried out using 9 cm diameter Petri dishes containing water and a little detergent. The traps were painted with one of five fluorescent colours or with yellow, blue, red, black or white gloss paints. The traps were spaced equally around the edge of a slowly rotating 2 m diameter turntable enclosed in a cage illuminated for 16 h/day and maintained at approximately 20°C and 55% RH. Carrot flies collected from emergence cages in the field were released into the test chamber and the numbers of flies caught were recorded daily.

Traps painted with the yellow gloss paint were twice as attractive as traps painted with yellow fluorescent paint. Both yellows were considerably more attractive than the other colours tested, the order of decreasing attractiveness being blaze = red = signal green > black > aurora pink = fire orange = blue > white.

2.4. Monitoring emergence of flies in the field

Yellow water-traps were used to monitor the emergence of carrot flies within the larger cages and in the adjacent open field. During peak activity of the two generations in 1981, approximately 40 and 20 flies/trap/day were caught in fluorescent yellow traps. Although peak numbers of flies were caught at approximately the same time, fewer flies were caught in the field than in the emergence cages and the peaks were less pronounced. In the field, traps were more effective in the crop than in a hedgerow 30 m away, where peak adult numbers occurred one week later. The ratio of male to female flies was 11:1 in the hedgerow traps and 1:10 in the crop traps (7). In addition, twice as many flies were caught in 9 cm diameter traps painted Bouton d'Or yellow (supplied by Dr E. Brunel, Rennes, France) than in the standard 16 cm diameter fluorescent yellow water-traps (8).

Table 1. Date and the number of soil day-degrees accumulated above 6°C for the first and second peaks of carrot fly activity during 1981 to 1985.

	1st Generation peak		2nd Generation peak	
	Calendar Date	Accumulated D° 6°C	Calendar Date	Accumulated D° 6°C
1981	18 May	275	17 Aug	1300
1982	12 May	250	11 Aug	1375
1983	20 May	216	24 Aug	1500
1984	15 May	260	17 Aug	1475
1985	23 May	265	6 Sept	1335

In 1982 and 1983, carrot flies were monitored again to determine the peaks of activity of the first and second generations so that the corresponding number of day-degrees (D°) accumulated at a soil depth of 6 cm from 1 February could be calculated. The results in Table 1 indicate that the first generations should be easy to predict from accumulated D° (Approx. 250D° above 6°C) and that, like the cabbage root fly, further biological information is required to improve the accuracy of the second generation forecast. Throughout all three years, peak numbers of flies and parasites were recorded at the same times.

3. DISCUSSION

Although the present Wellesbourne traps are suitable for monitoring populations of the carrot fly, sampling could be improved by making the traps more effective. Results from laboratory and field tests showed that the Bouton d'Or hue supplied by E. Brunel was approximately twice as attractive as fluorescent yellow. It was also somewhat surprising that although white traps are nearly as effective as yellow traps at capturing cabbage root flies (8), white was the least effective colour for attracting carrot flies. Future attempts to improve the visual aspects of the trap will concentrate on the various hues of yellow available as gloss paints. Particular notice will be taken not only of the numbers of carrot flies caught but also of the numbers and types of other species present in the traps. Recent results with the cabbage root fly have shown that certain hues of yellow, although not quite as effective as fluorescent at attracting cabbage root flies, are much more selective, and often trap only low numbers of beneficial syrphids. Another recent alteration that has made water-traps more effective for the cabbage root fly has been the addition of Campden tablets (Sodium metabisulphite/Potassium benzoate/Polyvinyl pyrrolidone - The Boots Co., Nottingham, England) to prevent bacteria attacking the drowned insects and subsequently discolouring the water in the traps. When two Campden tablets are added to the water of a standard 16 cm diameter trap, the water (1 litre) remains clear and the insects are preserved in good condition even when the traps are serviced only once a week. This improvement should also help to make water-traps more effective, since the traps can then retain their original hue throughout the trapping period.

In future, we do not intend to study the detailed biology of the carrot fly as sufficient of this work is already being undertaken at Cambridge (e.g. 9) and Rennes (e.g. 10). Instead we will concentrate on developing a model to forecast accurately the times of activity of this pest. If such a model is to be robust, it needs to be tested using data from as many situations as possible. We would like to co-operate with other members of the Group in this study, provided standard methods of sampling can be developed and agreed.

REFERENCES

1. FINCH, S. and COLLIER, ROSEMARY H. (1984). Development of methods for monitoring and forecasting the incidence of Delia radicum (brassicae) populations on brassicas. In: Agriculture. C.E.C. Programme on Integrated Control. Final Report 1979/1983. (Eds Cavalloro, R. & Piavaux, A.). Luxembourg: Office for Official Publications of the European Communities, pp. 287-302.

2. WHEATLEY, G.A. (1972). Effects of placement and distribution on the performance of granular formulations of insecticide for carrot fly control. Pestic. Sci., **3**, 811-22.

3. THOMPSON, A.R. and SUETT, D.L. (1982). Developments in insecticidal control of insect pests of U.K. vegetables. Scientific Hortic. **33**, 90-99.

4. ELLIS, P.R., FREEMAN, G.H. and HARDMAN, J.A. (1984). Differences in the relative resistance of two carrot cultivars to carrot fly attack over five seasons. Ann. appl. Biol. **10**, 557-564.

5. WHEATLEY, G.A. and FREEMAN, G.H. (1982). A method of using the proportions of undamaged carrots or parsnips to estimate the relative population densities of carrot fly (Psila rosae) larvae, and its practical applications. Ann. appl. Biol. **100**, 229-244.

6. FINCH, S. and SKINNER, G. (1980). Mortality of overwintering pupae of the cabbage root fly (Delia brassica). J. appl. Ecol., **17**, 657-665.

7. STEVENSON, A.B. (1980). Carrot fly. - Monitoring the seasonal development with traps. Rep. natn. Veg. Res. Stn for 1979, pp. 37-38.

8. FINCH, S. and SKINNER, G. (1974). Some factors affecting the efficiency of water-traps for capturing cabbage root flies. Ann. appl. Biol. **77**, 213-226.

9. JONES, O.T. and COAKER, T.H. (1977). Oriented responses of carrot fly larvae, Psila rosae, to plant odours, carbon dioxide and carrot root volatiles. Physiol. Entomol. 2, 189-197.

10. BRUNEL, E. and RABASSE, J.M. (1975). Influence de la forme et de la dimension de pièges a eau colorés en jaune sur les captures d'insectes dans une culture de carotte. Cas particulier des diptères. Ann. Zool. - Écol. anim. 7, 345-364.

Monitoring of the carrot rust fly, *Psila rosae*, for supervised control

H.den Ouden & J.Theunissen
Research Institute for Plant Protection, Wageningen, Netherlands

Summary

Monitoring of the carrot rust fly Psila rosae (L.) in open polder areas can be useful for developing supervised control in carrot growing. In this way, application of insecticides can be avoided on a considerable part of the carrot fields. Catches of Psila rosae were correlated with distances from high objects in the landscape and with percentages of infested carrots.

1. Introduction

Attacks by the carrot rust fly, Psila rosae, vary considerably with weather conditions and the character of the landscape. When the population density is high, soil treatment with insecticides at sowing time and/or spraying of the crop rarely can prevent serious damage and reduce it to a satisfactorily low level.

Often the low density of the first generation of P. rosae and the high numbers during late summer and autumn raise doubts about the necessity of soil treatment in spring. This treatment has consequences for the economy of the culture and for the population level of the natural enemies of the fly.

Spraying in late summer is determined by both the safety terms of the insecticide involved and the market prices of the carrots. Knowledge of the presence of P. rosae by monitoring and predictions of the degree of damage are important in decision making.

Finally, the threat of a developing resistance of the fly and that of bacteria which quickly can break down soil insecticides demand for caution towards the use of insecticides still available for this pest.

2. Materials and methods

Monitoring of the carrot rust fly can be performed either by a depot system or by trapping the flies in the field. The former gives rather good information about the time of emergence but the physiological conditions of the population present in the depot is rarely similar to that in the field (van 't Sant, 1961). Trapping by sticky traps according to the concept of Freuler et al. (1982) is more promising, but is too expensive for use in practice.

In our experiments we used an adapted version of the hanging yellow sticky trap of Berlinger (1980) originally consisting of a yellow painted lid of a polymer Petridish in which the sticky bottom of the dish is inserted. For our purpose this dish is fixed in varying numbers on a polyvinylchloride (PVC) pipe with a diameter of 16 mm and provided with

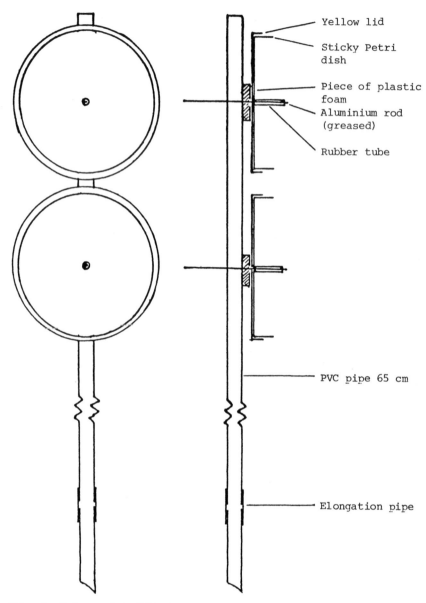

Fig. 1. Scheme of Modified Berlinger Trap to a Scale
        of 1 to 2

aluminum rods forced into just too narrow holes in the pipe (Fig. 1).
     The materials are very cheap. The polymer pipe with rods and yellow
dish lids can be used for many years and the sticky bottoms (in our
institute: recycled material from mycological experiments) can easily and
quickly be inspected in the field, eventually revised or sent up to an
inspector's address after covering them with lids. After inspection they
are discarded.
     The pipes were placed in the carrot field with the sticky plates just

162

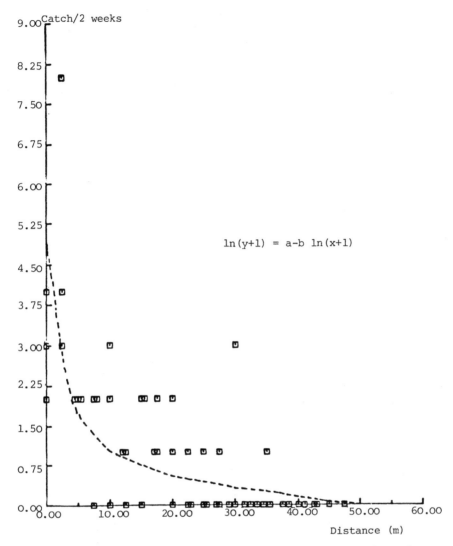

Fig. 2. The Relation Between the Numbers of P. rosae Caught and the
Distance of the Traps from Sheltering Objects in a Polder
Area. Catches at Distance Zero are Averages of 12-15
Trappings.

above the foliage of the crop. When the plants grew higher, the length of
the pipe was increased by an additional piece of PVC-pipe (25 cm).

In order to investigate the influence of the landscape on the presence and the flight activities of P. rosae the traps were placed in varying
numbers (10-20) in a parallel row to hedgerows or groups of trees etc. and
in a row perpendicular to these sheltering elements. The distance between
the traps was 2.5 m and the number of dishes per trap was two. In this way
an acceptable quantity of sticky surfaces was placed in the field and the
influence of the different sites could be observed particularly by increase and decrease of numbers of flies per trap in the perpendicular

% Infestation

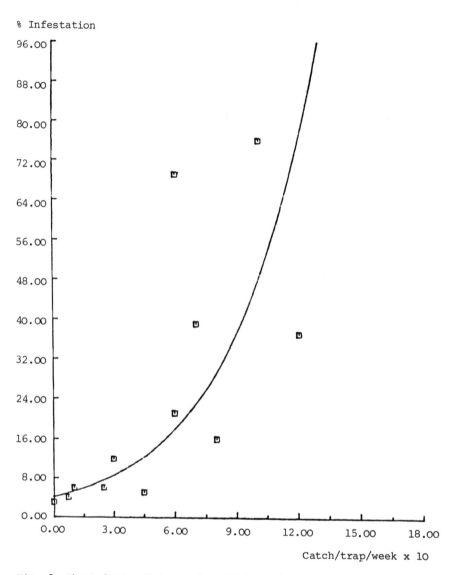

Fig. 3. The Relation Between the Catches of _Psila rosae_ and the
          Percentage Infested Carrots

rows. A more complete impression might have been obtained with a grid of
traps placed in the fields but then the amount of work per field would
have increased to a not-acceptable degree, particularly because the carrot
rust fly so often is absent in our open polder areas.

3. Observations

      In the years 1983-1985 observations have been made in open polders
and in more wooded areas. In these years particularly in early summer the
flights of P. rosae were poor. Mostly the second and/or third generation
showed better captures. Experiments in the polders where series of traps

164

Table 1. Catches in yellow sticky traps at different distances from high
objects in a polder landscape and percentages of infested carrots
in 1983, 1984 and 1985.

Average distance from object in meters	Week	Flies caught per trap	Percentage infested carrots	Date of sampling	Objects
5	36-37 1983	1.5	16	8 Oct.	Trees and shrubs
10		1	11		
30		0	2		
50		0	–		
5	35-38 1984	0.9	10	10 Oct.	Trees and shrubs
15		0.4	6		
20		0.2	4		
30		0	3		
5	35-39 1985	2.5	24	30 Sept.	Trees and shrubs
15		1.8	4		
30		1.0	3		
5	35-39 1985	2.8	45	30 Sept.	Road and ditch
15		–	16		
5	35-39 1985	0	0	30 Sept.	Wheat

were placed perpendicularly to each other as mentioned above provided
information on the occurrence of the adults of P. rosae on different sites
in the field. Fig. 2 shows the results of these experiments. The catches
in the series parallel to the hedgerows have been averaged and so the few
records on distance zero (actually a site circa 1 m into the carrot field)
have a weight that is 12-15 times higher than that of the records from the
single catches in the perpendicular series. A negative correlation between
the distance from the trees or shrubs and the catches is shown.

Table 1 shows the results of separate experiments in the years
1983-1985. A considerable decrease of the catches in the zone 0-20 m is
shown and a good correlation exists with the percentages of infestation in
the crops where samples of 100-200 carrots taken evenly over the different
zones have been collected.

A separate experiment done in a sheltered landscape with three double
dishes alternately directed North + South or East + West, showed an
average catch of 3.1 flies per trap in week 25. The traps were placed at
distances of 2.5 m from each other in two rows parallel to two hedgerows
perpendicular to each other at a distance of 40 m, forming a rectangle
with them. There was an even distribution of the flies over all the traps
in this landscape.

In week 25 + 26 twice as many flies were caught on the lowest of the
three dishes as on the upper two dishes (P ≪ 0.05). In September (weeks
34 + 35) when the foliage was fully grown catches on the three vertically
placed dishes did not differ any more.

4. Discussion

With the few available figures of percentages infested carrots a tentative curve could be fitted as is shown in Fig. 3. It might be necessary

to adapt the shape of the curve in future when more data become available.

The values are collected late August and early September. This is the season that growers are faced with a choice either to harvest the crop at short notice or to spray it and wait for a harvest two months later when the safety term will be expired.

Catches earlier in the season were generally lower. Therefore the economic feasibility of a soil treatment particularly in the cold polder areas is questionable and can only be interpreted as a safety insurance against a high number of flies of the first generation.

A promising approach might be late sowing in order to avoid soil treatment with insecticides against the first generation and consequently a late harvest which makes it easier to spray the crop more than two months earlier when tolerance levels in late summer might be exceeded. The threshold acceptable for a product with an economically sufficient quality should be assessed initially at a lower average figure than 0.1 fly per trap per week for 7% accepted infestation.

Consequently and according to Figure 3 a minimum of fifteen traps should be placed in the field at spots where the risk of carrot rust fly attack is high. The yellow traps should be placed very close to the top of the foliage particularly early in the season. This also has been shown by Freuler et al. (1982). More information should be obtained about the influence not only of hedgerows but also of wheat fields and ditches but also of dikes. The sheltering effect of maize fields, an increasingly important fodder crop in the Netherlands, that is maintained for a large part of autumn in the field will represent a threat just as serious for supervised control as the hedgerows described in Fig. 2.

Generally a zone of 20 m (representing one machine run) parallel to such objects should be treated only when the rest of the field is situated in an open windy space.

The trapping system described can be used either individually by e.g. a well trained farmer or in a team by establishing a scouting system e.g. similar to that organized in Switzerland. In both situations the sum of the time required for manufacturing and placing the traps, walking in the field and collecting the catchers will determine the costs.

Acknowledgements

Thanks are due to Dr. C.H. Booij for mathematical treatment of the data mentioned in Fig. 2 and Fig. 3.

References

1. Berlinger, M.J., 1980. A yellow sticky trap for white flies, Trialeurodes vaporarorium and Bemissia tabaci (Aleurodidae). Ent. Exp. & Appl. 27(1980): 98-102.
2. Esbjerg, P. Jörgensen, J. Nielsen, J.K., Philipsen, H., Zethner, O., og Ogaard, L., 1983. Integreret bekaempelse af skadedyr. Tidsshr. Planteavl. 87: 303-355.
3. Freuler, J., Fischer, S. and Bertuchoz, P., 1982. La mouche de la carotte, Psila rosae Fab. (Diptera, Psilidae) II Mise au point d'un piège. III Avertissement et seuil de tolerance. Revue Suisse Vitic. Arboric. Hortic. 14: 137-142, 275-279.
4. Sant, E. van 't, 1961. Levenswijze en bestrijding van de wortelvlieg (Psila rosae F.) in Nederland. Versl. Landbouwk. Onderz. nr. 67.1, Wageningen 1961, pp. 131.

Towards an effective and economic carrot fly trap: Progress report 1981-1985

E.Städler, F.Gfeller & K.Keller
Eidgenossische Forschungsanstalt, Wädenswil, Switzerland
H.Philipsen
Royal Veterinary and Agricultural University, Department of Zoology, Copenhagen, Denmark

The yellow plexiglass glue trap (20 x 20 cm, ICI 229) introduced by Freuler et al. (1982) replaced recently the less effective yellow water and glue traps as a monitoring tool for the carrot fly (Psila rosae). The only disadvantage of this trap is the fact that it is relatively expensive and therefore cannot be discarded after use. As a consequence, the traps have to be cleaned with solvent followed by retreatment with glue which is not without hazards and rather unpleasant. This prompted us to search for an alternative. The first trials using polypropylene with matching pigments to the plexiglass were unsuccessful, the polypropylene traps being far less attractive than the plexiglass. We speculated that the much higher transparency of the plexiglass could be the reason for its superiority. Consequently we replaced polypropylene by polystyrol with the pigments 9944, 9946, 9949 using the plastic mould developed for the fruit fly trap. The comparative experiments both in Denmark and Switzerland revealed no significant differences in the trap catches between the three pigments and the plexiglass. Since the pigment 9946 gave slightly higher catches, it was used for the commercial production of our carrot fly trap.

The identification of attractive components of the carrot leaf by Guerin et al. (1983) as field attractants for the carrot fly opened new ways to improve the attractivity and specificity of the plastic colour trap. However, in first experiments in the years 1981 and 1982 it proved to be difficult to reproduce the earlier positive results of Guerin et al. (1983). We suspected that the primarily used trans asarone was not very stable under field conditions and therefore a systematic study of the release rate of asarone from differently designed dispensors was studied. The amount of released asarone in the air above the dispensors was measured using the electroantennogram of female carrot flies. Based on these results, a dispensor has been designed consisting of a tube (metal or plastic) in which a filter paper impregnated with 100 mg of trans asarone, an antioxidant and a UV stabilizer is applied to the inner surface.

The results so far show that asarone is constantly released for more than a week in different seasons and localities of both countries. Trans asarone extracted from natural plant sources proved to be also active but was more expensive to be produced than the synthetic, pure compound. The ratio of the trap catches between traps with and without asarone was not constant during the two seasons of 1984 and 1985. Asarone was less active during the end of flight of the first and second generation. Future experiments will show if this effect is reproducible in the next years and what the possible explanation may be. In any case, our measurements showed that a change in the release rate of our dispensors is unlikely to be responsible.

We are in the process of correlating catches of the improved trap with the damage levels occurring in the host plants of harvest. As in the case of the earlier used yellow water and glue trap it will be necessary to accumulate the data of several years before threshold values in terms of trapped flies can be proposed.

REFERENCES

1. FREULER, J., FISCHER, S. and BERTUCHOZ, P. 1982. La mouche de la carotte, Psila rosae Fab. (Diptera, Psilidae) II. Mise au point d'un piège. Rev. Suisse Vitic. Arboric. Hortic. 14, 137-142.
2. GUERIN, P.M., STÄDLER, E. and BUSER, H.R. 1983. Identification of host plant attractants for the carrot fly, Psila rosae. J. Chem. Ecol. 9, 843-861.

Integrated pest management in Danish carrot fields: Monitoring carrot fly (*Psila rosae* F.) by means of yellow sticky traps*

H.Philipsen
Royal Veterinary and Agricultural University, Department of Zoology, Copenhagen, Denmark

Summary

Attack of carrot fly (Psila rosae) occur every year in carrot growing areas in Denmark. To minimize use of insecticides growers need to have information on risk of attack. For that purpose yellow sticky traps have been used to monitor the addult carrot fly. Earlier experiments have shown relationship between catches of flies, number of eggs round carrot plants and damage later in the season. But no damage threshold has been established.
To improve the value of yellow sticky traps efforts has been made to increase catches and to make the trapping system more selective. Some progress has been made, by sheltering the traps, but further development is needed.

Key words: Carrot fly, <u>Psila rosae</u> F., monitoring, yellow sticky trap, forecasting.

1. Introduction

In Denmark yellow sticky traps have been used for monitoring carrot fly (Psila rosae) since 1978. During this period damage level at harvest time has been estimated in many of the fields, where traps have been used. It is not to be expected that there will be a direct correlation between numbers of flies caught and the damage level at harvest. However, it has been shown that, generally more carrots are attacked on sites where many flies have been caught (Esbjerg et al. 1983).

An uneven distribution of flies in individual fields results in similarly uneven distribution of damage in these fields. One consequence of this is, that it is unlikely, that a precise damage threshold can be established. This difficulty is further stressed by the fact, that considerable

* The present paper includes results from several investigations which have all been supported financially by the Council for Agricultural and Veterinary Research. The first part (incl.1982) formed part of the project "Integrated control of insect pests", which used carrots as model crop and involved cooperation with P. Esbjerg, Danish Research centre for Plant Protection, Lyngby and J.K. Nielsen, O. Zethner and L. Øgaard at the Royal Veterinary and Agricultural University, Department of Zoology.

damage has occured at sites with low catches. Nevertheless, yellow sticky traps give some useful information on flight activity. Zero catches should now lead to negative warning, while increasing catches can give reason for consideration of insecticide treatment.

2. Background

During an earlier project the usefulness of the sticky traps was analysed especially concerning relations between catch and egglaying, but also concerning differences in activity at different sites.

It was observed, that during the main period of flight there was good agreement between changes in population size, measured by flies caught on traps and measured by numbers of eggs sampled in the field immediately round carrot plants (Esbjerg et al. 1983). However, this was not the case at the beginning of the first flight period, even though it was shown, that the flies caught on sticky traps contained mature eggs and at the same time many eggs were found around wild umbelliferous plants (Esbjerg et al. 1983).

To investigate the influence of crop development on egg-laying, plots of well developed carrot plants were established by early sowing and protection of the seedlings with polythene sheets. From fig. 1 it can be seen, that the presence of well developed plants at the beginning of the flight period establishes a situation, where there is a good agreement between catches and presence of eggs around carrot plants. It was concluded, that carrot plants under field conditions should be developed beyound the two to three leaf stage to be recognised or accepted by the carrot fly females for egglaying.

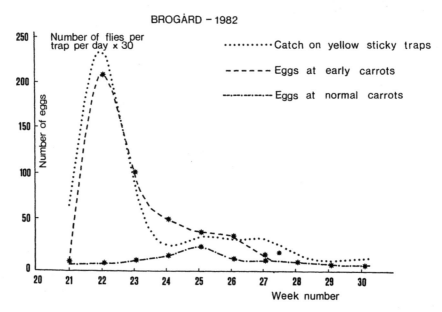

Fig. 1. Catch of carrot flies on yellow sticky traps shown together with number of eggs collected around early sown and protected carrots and around normally sown carrots.
Week numbers refer to periods for accumulated catches and egg laying, e.g. week number 21 refers to trap catches in the period May 17th - 24th and egg sampling May 24th. (Esbjerg et al. 1983)

The need for and value of monitoring is stressed by the fact, that flies do not occur simuntaniously at all sites. Fig. 2. shows the number of flies caught at three different sites in 1980. It is noticeable that there is a difference of almost a month between flight peaks at Rønnely and Brogaard, the two sites being appr. 8 km apart. The distance between Brogaard and Strandholm is 1.5 km. The differences in peaks of activity on the three sites are due to differences in soil-temperature and -conditions.

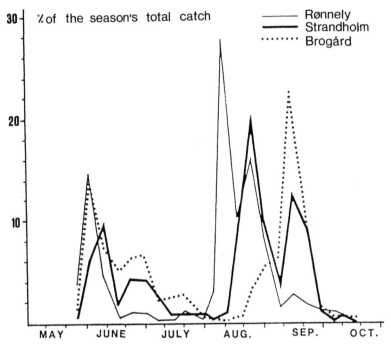

Fig. 2. Relative distribution of carrot fly at three localities in a carrot growing area in NW-Zeeland, 1980 (Philipsen, 1983)

Numbers of trapped females have been used when investigating the relationship between egg-laying and flight-activity, because this is a better expression for damage potential than total number of flies (males and females). For practical purposes it would be less time-consuming merely to count all trapped carrot flies and hence establish the risk level.

To find an appropriate correction factor proportions of males in the catches were estimated. Fig. 3 shows the proportion of males in catches throughout the season. It can clearly be seen, that males occur in lower proportions in connection with the first flight period than at the end of the season, where sex ratio is approximately 1:1.

The main conclusion concerning further work with yellow sticky traps was, that efforts should be made to obtain bigger catches. This could be done either by making traps more effective or by using more traps at the individual locality, thus making the basis for decission more sound (Philipsen, 1983).

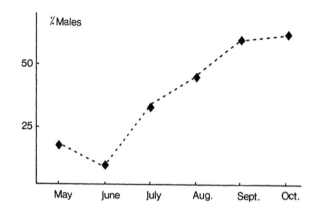

Fig. 3. Percentage of male carrot flies caught on
yellow sticky traps at Strandholm, based on
summarized monthly catches from 1983 & 1984.

3. Trap efficiency

In the following some results from the efforts of improving catching
results will be reported.

Three lines were followed

A) To make the traps more conspicuous in the field.
B) To incorporate the flies' preference for sheltered positions in the
   trapping procedure.
C) To use attractants in combination with the traps.

The latter part was carried out as a joint project in coorporation with
Erich Städler, Wädenswil, and will be reported elsewhere.

3.1. More conspicuous traps

The surroundings of traps are known to influence trapping efficiency.
To test a possible positive influence of bare soil as "background" for
yellow sticky traps trapping efficiency was registered on traps placed in
small plots (1.2 x 1.2 m), where all plants were removed. Catches were compared to catches from traps placed in the same carrot fields, with carrot
foliage as "background". From fig. 4. it can be seen that a background of
bare soil does not increase catches, but has a negative effect on catch
level, hence traps placed on bare soil are not more conspicuous to carrot
flies. The practical implications of this information are, that traps
should always be placed in parts of the trapping area, where there is a
good, even stand of carrots.

3.2. Influence of shelter

It has, on some occations, been observed, that catches could be influenced by tall weeds, bolters or other obstacles near the traps. To test the

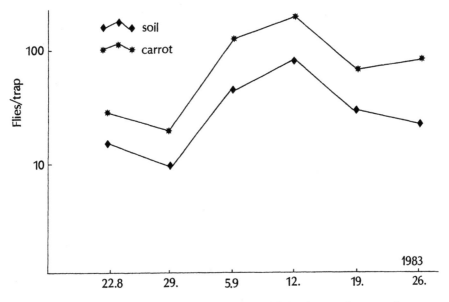

Fig. 4. Catch of carrot fly on yellow sticky traps in carrot
fields or in plots of bare soil (1.2m x 1.2m) in the
same fields. Each point represents mean catch from
3 traps.

relevance of these observations yellow sticky traps were placed inside and
immediately outside net-cylinders, (mesh 5 x 5 mm) see fig 5. Catch level
from these traps was compared to catch level on traps placed in the usual
manner - on iron stakes immediately above crop-level.

Preliminary results, fig. 6., indicate larger catches on traps adjacent
to the netting and smaler catches on traps inside the netting. However, there seems to be one great advantage of placing the traps inside the netting.
Traps become more selective in the sense, that they catch considerably fewer
"other insects", the proportion of carrot flies thus becomes greater, making
counts far easier.

These observations need further experiments, in order to establish
whether selection of a more appropriate mesh will further improve these
advantages and whether the effect is consistent throughout the whole growing
season. In addition it is important to insure a final trap design, that can
be used in practice.

4. Yellow sticky traps in supervised and integrated control

Use of yellow sticky traps should be regarded as a first and obvious
step in order to minimize use of insecticides in carrot growing, because it
gives possibilities of negative warning, so routine spraying can be avoided.
In the areas where the aforementioned trials were carried out, the informations on abundance of carrot flies resulted in decrease in numbers of treatments and the system was recommended to be used on a wider scale.

From 1984 traps have become more widely used, 14 and 24 sites in 1984
and 1985 respectively. The disposable plastic sticky traps (REBELL®) are
distributed by The Danish Research Centre for Plant Protection, Lyngby who
also collect results weekly and inform Extension Service about the general

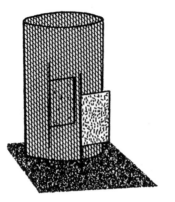

Fig. 5. Plastic net-cylinder
        (h: 1m d: 0,55m) with
        yellow sticky trap in-
        side and just outside
        the cylinder.

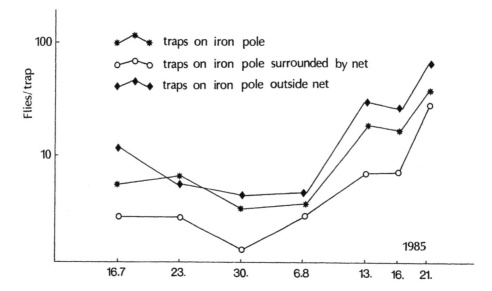

Fig. 6. Catch of carrot fly on yellow sticky traps in
        carrot field and on traps placed inside or
        just outside a net-cylinder in the same field.
        Each point represents mean catch from 2 traps.

situation. Locally, the Extension Servic know the situation from their own
traps.

It is far too early to draw conclusions, but one result is that, two
years trapping have shown that fly populations are at a very low level at
many localities. This has given the basis for avoidance of routine treatments.

Use of yellow sticky traps can also be regarded as a step towards integrated pest management in carrot crops. The next steps under Danish conditions, concerning carrot fly, could be the use of knowledge concerning
the influence of attacks of fungi (Entomophthorales) on the potential population development (Eilenberg, 1983), and avoidance of carrot growing in
sheltered and therefore high-risk areas.

REFERENCES

1. EILENBERG, JØRGEN (1983). Entomophthorales on the carrot fly (Psila rosae F.). Possibilities of using Entomophthorales for biological control of pests. - Tidsskrift for Planteavl 87:399-406. Danish, English summary.
2. ESBJERG, P., JØRGENSEN, J., NIELSEN, J.K., PHILIPSEN, H., ZETHNER, O. and ØGAARD, L. (1983). Integrated control of insects with carrots, the carrot fly (Psila rosae) and the turnip moth (Agrotis segetum) as crop-pest model.-Tidsskrift for Planteavl 87:303-355. Danish, English summary.
3. PHILIPSEN, HOLGER (1983). The use of yellow sticky traps to estimate the risk of attack by carrot fly (Psila rosae F.). - Tidsskrift for Planteavl 87:389-397. Danish, English summary.

Integrated pest management in Danish carrot fields: Monitoring of the turnip moth (*Agrotis segetum* Schiff., Lepidoptera: Noctuidae)*

P.Esbjerg

Danish Research Centre for Plant Protection, Institute of Plant Pathology, Zoology Department, Lyngby, Denmark

Summary

Cutworm attacks in Denmark vary greatly from year to year. Hence forecasting is important as background for and economic handling of this pest problem. A monitoring system based on sex traps has been developed through the years to obtain the necessary information about the pest. Despite the fact that the catches of the turnip moth have not corresponded to the damage, it has been possible to issue reasonable forecasts. These forecasts were based on catch and weather conditions of importance for the development and mortality of the cutworms. The system has been improved, but further development is desirable. For this, however, more precise information on development and mortality factors have to be obtained experimentally in order to make a more realistic estimate of the change of populations from moths until 3rd instar larvae.

The pest status of the cutworm (<u>Agrotis</u> <u>segetum</u>) in Denmark varies from the sporadic level (as for instance in 1985) to an almost disastrous level (1976). The considerable fluctuations which appear from the records (Stapel, 1977) have been analysed by Mikkelsen and Esbjerg (1981). By linear regression analysis it was seen that certain weather trends play a major role for the fluctuations. A simple model was worked out, but the authors stressed that the model should only be regarded as a valuable supplement to the local monitoring which was started during the late seventies. As also mentioned by Emmet (1983), a monitoring system is essential if unnecessary and wrongly timed insecticide applications as well as a repetition of the losses incurred in 1976 are to be avoided.

The early start of the turnip moth monitoring system was based on an experimental sex trap using virgin females as bait. The reliability and amount of catch was, however, increased by means of an improved trap design (Esbjerg et al., 1981). Besides, the development of a synthetic pheromone in Switzerland - a Danish/French/Hungarian cooperation (Arn et al., 1983) led to a very important standardisation of the trapping. Recently, further

*
The present paper includes results from several investigations which have all been supported financially by the Council for Agricultural and Veterinary Research. The first part (incl. 1982) formed part of the project 'Integrated control of insect pests', which used carrots as model crop and involved cooperation with J.K. Nielsen, H. Philipsen, O. Zethner and L. Øgaard at the Zoological Institute of the Royal Veterinary and Agricultural University in Copenhagen.

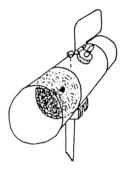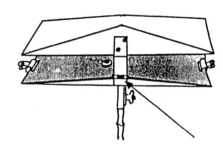

Fig. 1. Left: The experimental trap design used in the late seventies.
       Right: The trap design used at present for the monitoring of turnip
              moths in Denmark. The arrow points to a hinge. On the other
              side of the trap (which cannot be seen on this figure) there
              is snap opener, so that the whole trap roof may be folded
              back.

practical improvement has been obtained by minor but important changes in
the trap design (Figure 1, Esbjerg et al., 1981).

THE PRESENT MONITORING SYSTEM

        The basic tool is the trap shown on Figure 1 equipped with a 1:1:1
blend of Z-5, 10 Ac; Z-7, 12 Ac; and 29, 14 Ac; according to Arn et al.
(1983). The rubber dispenser (serum bottle cap No. 90142, Auer Bittman
Soulie Zürich) is loaded with 100 mg of the blend. At each monitoring
locality, three traps are set up forming an equilateral triangle with a side
of 50 m. Growers are instructed to set up the traps within a field away from
front hedges, banks or other major obstacles. The pheromone dispenser is
changed after six weeks, and the sticky cardboard in the trap is changed if
more than 25 moths are caught. Normally the catch is counted twice a week,
but during peak flight the countings may occasionally take place every day
to avoid overloading of the trap bottom.
        Four permanent research localities form the backbone of the system,
but on top of this, temporary localities are set up at farmers' request and
payment. Traps are rented from the Plant Protection Centre at a price of
approx. 500 Dkr for three. Included in this price are dispensers, sticky
bottoms and forecasts/warnings. The latter are mailed to the farmers and
their respective extension officers.
        Since the start of the monitoring system the number of trapping
localities has increased as follows:

Year	1977	1978	1979	1980	1981	1982	1983	1984	1985
Number of localities	11	12	13	13	13	17	19	33	49

Trap and bait:		
Experimental	Experimental pipe type	Design as shown on Figure 1

On Figure 2 the trapping localities used in 1985 are mapped.

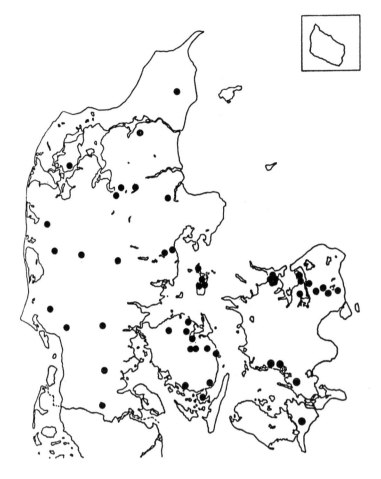

Fig. 2. The Geographical Distribution of Turnip Moth Trapping Localities in
        Denmark 1985

CATCH

    To give an impression of the catches on an annual basis Table 1 shows
total catches on 12 localities during 1983, 1984 and 1985. The figures from
Onsbjerg, Samsø, in particular, show how much the catches may vary from one
year to another. Besides, the distribution of catches and the length of the
catching period varies considerably, as appears in Figure 3. This figure
shows graphs of catches from the same locality (Samsø) throughout the years.
As the curves have been drawn through rolling means of catches from five
nights, the graph is somewhat smoothed in comparison with curves drawn
through actual catches. The latter will be much more irregular because of
changing activity, while the smoothed curves may give a better impression of
the moth populations.

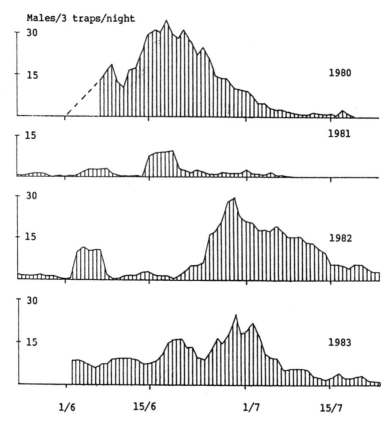

Males/3 traps/night

1980

1981

1982

1983

1/6   15/6   1/7   15/7

Fig. 3. Graphical Presentations of Sex Trap Catches of the Turnip Moth
During Six Subsequent Years on the Island of Samsø.
The curves were drawn through means of five nights in order to
smooth the variation between nights and rather illustrate the
variations in length, time distribution and peak height of the catch
in different years. The dotted curve part in 1980 is based on
estimation from sex trap catches with virgin females as the bait.

Figure 4 shows another aspect of turnip moth catching. This is the
possible influence of major obstacles in the landscape. Thus the two curves
represent catches with two sets of traps only 600-700 m apart, but separated
by buildings, trees etc. If no obstacles of this nature are present, catches
at that distance will normally be similar. Therefore many farmers with
neighbouring fields without major separating obstacles may share one set of
traps. Besides, this, the zig-zag appearance of the curve for Arslev H
(4th-18th July) shows a 'mild' case of insufficient removal of the sticky
cardboard bottoms. The catch figures are proportionally too low when the
bottoms are slightly overloaded with caught moths.

CATCH INTERPRETATION

The sex trap catches of turnip moths are used as the primary
background for forecasting of cutworm attacks so early that treatment can be
carried out before any great proportion of the cutworms enter the 3rd

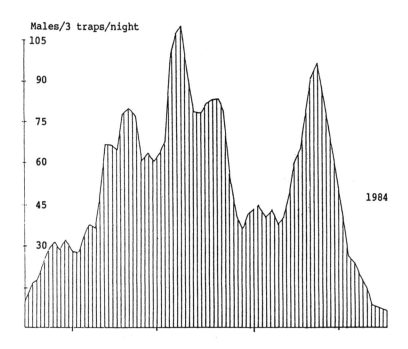

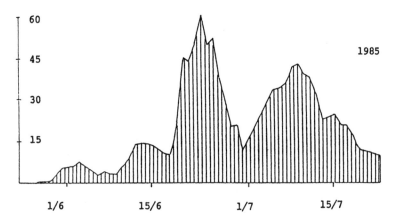

Fig. 3. (cont.)

instar. It has only happened once (1981, cf. Figure 4) that trap catches were so low everywhere that negative warning could have been issued solely on that background. During the last few years, a catch of five moths per three traps per night has been used as a provisional threshold below which there should be no risk at all. Twenty five males per three traps per night has been used as an additional threshold, above which increased attention is necessary and damage may be likely, unless a fairly large amount of rain spread over a number of days (e.g. 50 mm spread on 10 precipitation days)

TABLE 1 <u>Annual Catches of Agrotis segetum Males in Sex Traps at Selected Localities for Three Years.</u>
The figure for Ulfborg is shown as a substitute for Thisted. On these localities, soils are fairly similar, and so was the weather in 1984 and 1985. Like Studsgard they are both situated in the western part of Jutland, where catches were in general increasing from 1984 to 1985.

Trapping locations	1983	Total catch 1984	1985
Virumgård, Lyngby	411	804	800
Bonderup (Frederikssund)	537	2369	1630
Stubberupholm, Lammefjorden	794	2647	1256
Broby (at Odense)	-	1192	711
Arslev	975	2255	1464
Tåsinge	417	693	408
Onsbjerg, Samsø	500	3380	1178
Ulfborg	-	-	1416
Thisted	794	791	-
Godthåb, Skanderborg	334	772	668
Studsgård (Herning)	212	559	1833
St. Jyndevad	730	837	738

increases mortality before the date of treatment is reached. As indicated by this, the forecasts are based on catches as well as on weather records. However, they are also differentiated in accordance with different soil types and crops. The most important crops ranked after sensitivity are: Red beets (most sensitive), carrots, leeks, onions and potatoes.

The most important limiting factor is the mortality of 1st and 2nd instars, which may be caused by moist soil, as shown experimentally (Esbjerg et al., in preparation; Esbjerg, in preparation). These experimental results and the statistical results of Mikkelsen and Esbjerg (1981) are practically ruled out in the following way.

The mortality increases (and the risk diminishes) with the amount of precipitation and the number of precipitation days. As the influence of precipitation is not direct, but through the consequent soil moisture, the risk is lower on clayey than on sandy soils if the summer is not extremely dry as in 1976.

The influence of temperature has primarily been taken into account in relation to the timing of treatments. However, low temperature during the maximal occurrence of 1st and 2nd instar cutworms has also been regarded as reason for increased mortality if occurring synchronously with moist soil. The reason for this is the strong temperature dependence of the duration of 1st and 2nd instars: in the range of one month at 15°C and 1/3 or less at 25°C (Esbjerg, unpublished data), 1st-2nd instars also seem to form the part of the life cycle which is most sensitive to high soil moisture (Esbjerg, unpublished results).

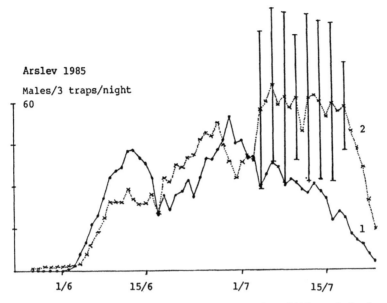

Fig. 4. Catch Curves of _Agrotis segetum_ at Two Localities at Arslev Only About 500 m Apart From Each Other, but with Buildings and Rows of Rather Large Trees in Between.
The curves are drawn through means of five nights, and despite the size of standard deviation shown on part of the 2-curve, the catches are significantly different (99% level, Man-Witney U-test (Siegel, 1956) - at any rate in the period 4th-18th July. The rather regular zig-zag appearance is in agreement with too late changes (overloading) of sticky trap bottoms giving a decreased catch every other time.

FORECAST IN PRACTICE

Even though the sex trap catches and the knowledge of the influence of the weather has given a very useful background, there has also been quite an element of trial and error in the forecastings. In view of this, the verbal forecasts have always tended to advise slight overtreatment rather than lose the farmers' confidence.

To check the trial and error element, a variable number of samples from both treated and untreated areas have been investigated for damage. Table 2 shows a summary of results together with the mainline of advice included in the forecasts. As the percentages of damaged red beets mainly refer to Samsø, these may be conferred to the catch curves on Figure 4. In addition to the information from Table 2, it should be mentioned that no farmers who treated as recommended in 1982, 1983 and 1984 had more than 2% damaged roots, and quite a large number had a level of 0 or 0.5%.

BENEFITS OF THE FORECASTING - FURTHER DEVELOPMENT

The fertility of the turnip moth is very high (about 800 per female, Weissmann and Podmanicka, 1971) and the mortality is highest before the damaging 4th, 5th and 6th instars. Therefore, a direct relationship between sex trap catches and the final damage level would be rather unlikely. This

183

TABLE 2. Summary of Forecasts From 6 Years and the Main Reasons for the Forecasts as well as Damage in Untreated Parts of Fields with Carrots or Red Beets. (The red beets mainly refer to Samsø, cf. Figure 3). Although it is based on a limited amount of samples, the last column of figures gives a clear impression of the great varying cutworm densities.

Year	Major reason for content	Summarised main content of forecast	% damaged root in samples from untreated field parts		Roughly estimated number of 6th instar cutworms per m² based on limited sampling
			Carrots	Red beets	
1980	Excessive rainfall	Do not treat	0	0	-
1981	Low catch and strong rainfall	Do not treat	0	0	0.02
1982	Low rainfall in July in regions with light soils	1 treatment recommended in particular crops in 2 distinct regions	10	50	-
1983	Dry and warm weather throughout July	2 (-3) treatments recommended in selected areas with certain catches and particular crops	10-20	25-70	10-25
1984	High catches and in spots, temporary lack of precipitation	1-2 (or 3) treatments in widespread areas depending on crop, soil and precipitation	0-20	3-50	1-5
1985	Medium catches and regional lack of precipitation during first half of July	0-2 treatments depending on catch, crop, soil and precipitation	0-1	0-2	0.1

is fully confirmed by the almost complete lack of correlation between the catches as illustrated by Figure 3 and the damage shown in Table 2. However, the issued forecasts have been reasonably good, considering that:

1) <u>No</u> grower following the advice given in forecasts has had any economically significant loss due to damage.
2) The number of pesticide treatments has been markedly reduced in proportion to the number which would have been used following a routine programme (2 or 3 fixed treatments per season).
3) According to control counts of damage in untreated field parts, the great majority of treatments have been justified, and only in one year (1985) there was a general recommendation of unnecessary treatments.

The last point may to some extent be explained by the experimental results on influence of temperature on cutworms (Esbjerg, unpublished) obtained later than the forecast and by special weather conditions. Thus delay of development and in particular mortality due to low soil temperature seem to have been seriously underestimated, which led to the first recommendation of treatment. In connection with the second treatment this was only a factor of minor importance. The start of heavy rains just around the time for the second treatment played a much more important role. This rainy period, which went on for more than one month, undoubtedly led to a very high mortality among young cutworms. Such a mortality was confirmed by the finding of a considerable number of red beets (25-30%) with tiny scars on the roots made by 3rd instar cutworms, which had just started their gnawing.

FUTURE IMPROVEMENT

Despite the general success of cutworm monitoring in Denmark based on sex trap catches in 1985, the results show that further improvement is desirable. This may be obtained primarily by improving the background for the interpretation of the catches.

Further details on development rates and mortality in relation to soil moisture and temperature are particularly important.

It would also be preferable to elucidate the variations in catch in relation to weather conditions during the night. Then a partial catch curve could be corrected for this variation and give a reasonably realistic impression of the population based on the amount of catches and the slope of a corrected curve. Such a possibility may be important if more farmers want to enter the monitoring system later in the season because of weather conditions favourable for cutworms.

The next improvement will be the development of a simple dynamic simulation model based on catch instead of the report-based index used in the statistical model of Mikkelsen and Esbjerg (1981). Besides, statistical results from analyses of mean monthly precipitation figures etc. should be replaced by information derived from experimental results on development and mortality. In this way, a new model might also have the operational advantage of making gradual updating possible. At present it is only possible to update the model from England produced by Bowden et al. (1983). However, this model does not give reasonable results with Danish cutworms, perhaps because of the use of development rates from cutworms living under very different geographical conditions.

REFERENCES

ARN, H., ESBJERG, P., TOTH, M., SCOCS, G., GUERIN, P. and RAUSCHER, S.,1983. Field attraction of Agrotis segetum males in four European countries to mixtures of three homologous acetates. Journ. Chem. Ecol. 9(2), 267-276.

BOWDEN, J., COCHRANE, J., EMMETT, B.J., MINALL, T.E. and SHERLOCK, P.L., 1983. A survey of cutworm attacks in England and Wales, and a descriptive population model for Agrotis segetum (Lepidoptera: Noctuidae).

EMMET, B.J., 1983. Pheromones in UK farm pest control. EEC Experts' Meeting, Parma, Oct. 1983, 47-57.

ESBJERG, P., NIELSEN, J.K. and ZETHNER, O., 1982. Influence of trap design on catch of the turnip moth (Agrotis segetum, Lep. Noctuidae) males in sex traps baited with virgin females. Les Médiateurs Chimiques, Versailles, 16-20 Nov. 1981, Ed. INRA Publ., 1982 (Les colloques de l'INRA, 7)

ESBJERG, P., NIELSEN, J.K., PHILIPSEN, H., ZETHNER, O. and ØGARD, L., in preparation. Soil moisture as mortality factor for cutworms (Agrotis segetum, Lep. Noctuidae).

ESBJERG, P., in preparation. Semifield experiments on cutworm (Agrotis segetum Lep. Noctuidae) mortality in relation to precipitation and soil type.

MIDDELSEN, S. and ESBJERG, P., 1981. The influence of climatic factors on cutworm (Agrotis segetum) attack level, investigations by means of linear regression models. Beretn. nr. 1561, Tidsskr. Planteavl 85, 291-301.

SIEGEL, S., 1956. Nonparametric statistics, McGraw-Hill, London.

STAPEL, CHR., 1977. Den økonomiske betydning af plantesygdommenes og skadedyrenes bekaempelse i landbruget. Ugeskr. f. Agron., Hort., Forst. og lic. 122, 735-746.

WEISMANN, L. and PODMANICKA, 1971. Einfluss der Temperatur und Nahrung auf die Entwicklung der Wintersaateule Scotia Segetum (Denis et Schiffermüller). Biologische Arbeiten, Wissenschaftliche Kollegien für allgemeine und spezielle Biologie der slowakischen Akademis der Wissenschaften XVII/8, 1-76.

Control of *Heterodera carotae* Jones (1950): Effectiveness of available methods and prospects

M.Bossis

INRA, Laboratoire de Recherche de la chaire de Zoologie, Domaine de la Motte-au-Vicomte, Le Rheu, France

Summary

Several nematodes, as well as insects and fungi, can cause damage to carrots. One species, Heterodera carotae, is prevalent in areas of intensive carrot production in parts of western France. It also occurs in some other countries. This paper discusses the effectiveness of control measures and indicates prospects. The use of non-host plants is difficult owing to the high specificity of the parasite and a very low tolerance limit (< 1 larva/g). A five year rotation (or ten year for heavy infestations) is recommended. Fumigant treatments are effective, but recolonisation occurs during the growth of the host crop and populations sometimes exceed their initial levels. Treatments must be applied before each carrot crop is sown. Biological control with Paecilomyces lilacinus does not protect crops effectively, even though the fungus colonises the soil. The long term effects remain to be tested. An in vitro test is proposed for breeding for resistance.

1. INTRODUCTION

A number of field carrot pathogens are capable of causing a drop in yield or large scale losses during storage. As well as insects (flies and aphids) and various fungal diseases on foliage (Alternaria dauci, Plasmopara nivea, Septoria carotae, etc.) or roots (Rhizoctonia violacea, Phytophthora megasperma, etc.), several nematodes can cause damage (4). Among the more common of these, the meloidogynes (incognita, javanica, hapla) and Heterodera carotae appear to be most harmful. The latter is very widespread in most European countries and was recently discovered in the United States (Golden, personal communication). The regions of intensive cultivation in western France (Manche, Finistère, Morbihan, Loire-Atlantique) are particularly heavily infested: Oudinet (22) reports that H. carotae is widespread in Manche and that 60% of the cultivated area is infested in the most badly affected zones, but that the effect of other nematodes is negligible. In this Department, analyses performed on 120 plots throughout the coastal areas where carrots are most frequently grown showed the parasite to be present in two-thirds of cases (Table 1), the heaviest infestation level being 230 cysts/100 g soil, or approximately 100 larvae/g. It is also found in several other regions of France such as Bordelais, Lyonnais, Val de Loire and the Paris area, although little is known about its distribution.

Very heavy infestation (> 50 larvae/g soil) virtually wipes out carrot crops. Moreover, according to both Ambrogioni and Marinari Palmisano (1) and

TABLE 1. Results of Analyses of 120 Plots
(Département de la Manche).
H. carotae Cyst Infestation
Levels in 300 g Soil

Number of cysts	% of plots
> 100	29
50-99	12
10-49	17
1-9	9
0	33

Greco and Brandonisio (11), the harmfulness threshold is less than
1 larvae/g, i.e. extremely low. Producers in infested areas therefore have
to take precautions against the parasite. We will use the biological data at
our disposal to investigate current strategies for maintaining this nematode
at harmless levels and the longer term prospects.

2. CONTROL STRATEGIES

2.1 Use of Non-Host Plants

    H. carotae is a highly specific nematode whose range of hosts is
restricted to a few umbellifers (16-23). The larvae contained in the cysts
require carrot root exudates in order to hatch (24). Aubert (2) has shown
that some umbellifers are able to stimulate the emergence of larvae but that
no cultivated crop other than the carrot provokes large scale hatching. This
rules out the use of any trap crop except the carrot itself; the risk of
regrowth is too high for this technique to be considered, and it has not
been tested. The nematode may produce two generations per year under normal
growing conditions, particularly with carrots produced for storage.
    When no carrots were present, we observed a 33% decrease in one year
in western France (3); however, a number of edaphic and climatic factors may
affect this natural mortality. In any event, assuming a one-third reduction
each year, it would take at least 10 years for the highest populations to
drop below the harmfulness threshold, and five years even at low infestation
levels. A rotation interval of this length has very little chance of being
adopted by farmers. For 'sand carrot' production in particular, there are
relatively few areas of cultivation with light soils and climates mild
enough to allow overwintering in the field, and long rotation intervals are
out of the question.

2.2 Chemical Treatments

2.2.1 Fumigants

    The effectiveness of fumigant treatments against H. carotae needs no
further demonstration. Oudinet et al. (20), Oudinet (21) and Greco and
Lamberti (9) have noted very sizeable improvements in yield. Application of
dichloropropane-dichloropropene (DD) or dibromoethane reduces populations to
levels at which very satisfactory harvests are possible (Table 2). However,
the population recovers in a single crop cycle and may even exceed the
original level by a substantial margin (3). The treatment therefore has to
be repeated before each new crop is sown.

TABLE 2. Effects on Yield of Various Fumigant Treatments
         Against H. carotae

Researcher	Product	Dose	% increase in saleable yield
Oudinet (1966 experiment)	DD	400	341
Oudinet (1966 experiment)	Dibromo-ethane	150	334
Greco (1973 experiment)	DD	400	2400
Bossis (1982 experiment)	DD	315	603
Bossis (1984 experiment)	Dibromo-ethane	220	444

2.2.2 Other products

Results are generally far inferior to those obtained with fumigants
(8).

2.3 Biological Auxiliaries

Phytophagous and mycophagous nematode antagonists have already been
tried several times. Kerry (17) lists the main fungal species used. The
farmer may use a nematophagous fungus, Artrobotris irregularis; an approved
strain is marketed under the name of 'Royal 350' (5, 6). Although this
fungus is effective against meloidogynes, it is relatively useless against
Heteroderinae larvae. Another fungus, Paecilomyces lilacinus, may live on
Meloidogyne incognita acrita and Globodera pallida (13). Applications have
yielded good results in the field (14) and beneficial effects have been
noted over several years (15). P. lilacinus attacks only very young eggs in
the primary stages of embryogenesis; it cannot therefore be expected to
affect the contents of old cysts whose eggs have completed their embryonic
development and are responsible for infestations of young carrot sowings.
This fungus, grown on a PDA medium, has been used in preliminary tests both
in containers and in the field in an attempt to limit populations of H.
carotae. In the field, a comparison was made with a 220 l/ha dibromoethane
treatment.

Table 3 shows the initial populations of H. carotae in the container
experiment; the populations in the field were 14-123 cysts/100 g soil, i.e.
10-58 larvae/g.

Tables 4 and 5 show the levels of contamination by P. lilacinus
observed in the containers and in the field 15 weeks after inoculation. The
fungus was very well established even at the lowest dose. A slight
colonisation of the control was probably due to wind dissemination of the
mycelium. Initial inoculation levels were the same for both types of
experiment.

189

TABLE 3. Populations of H. carotae (Cysts and Larvae)
Before Inoculation of Containers with P. lilacinus

Dose in g per 100 l soil	Cysts/100 g	Larvae/g
0.2	199	34.6
2	179	32.4
20	167	29.5
Control	203	37.6

TABLE 4. Average Number of P. lilacinus
Propagules Found 15 Weeks After
Inoculation of Containers

Inoculant dose	Number of propagules/g soil
0.2 g/100 l	10^3
2 g/100 l	5.67×10^3
20 g/100 l	3.23×10^4
Control	0

TABLE 5. Average Number of P. lilacinus
Propagules per g soil Found
15 Weeks After Field Inoculation

Inoculant dose	Number of propagules/g soil
0.2 g/m^2	10^3
2 g/m^2	9.6×10^3
20 g/m^2	1.3×10^5
Control	1.3×10^3
Dibromoethane	0

Contamination of females, cysts and egg masses was, however, very poor. Only the strongest inoculation doses contaminated H. carotae in a few cases (Tables 6 and 7).

TABLE 6. Contamination of H. carotae Females and Egg Masses 15 Weeks After
Inoculation of Containers With P. lilacinus
(dose: 20 g/100 l)

	Numbers	Number of contaminations by P. lilacinus
Young white females	100	0
White females with egg sacs	80	0
Cysts with egg sacs	35	0
Egg sacs	20	2

TABLE 7. Contamination of H. carotae Females and Cysts 15 Weeks After
Field Inoculation With P. lilacinus
(dose: 20 g/m^2)

	Numbers	Number of contaminations
White females with egg sacs	100	6
Cysts with egg sacs	100	0

Table 8 compares the yields obtained in the field trial. In the presence of a relatively high H. carotae population, the fungus proved totally ineffective and incapable of protecting the crop.

TABLE 8. Average Yield of Saleable Carrots After Application
of P. lilacinus or Dibromoethane

	Yield in kg per 6 linear metres
Control	5.3 a*
P. lilacinus (20 g/m^2)	4.8 a
Dibromoethane (220 l/ha)	23.5 b

n = 20
* a and b differ significantly at P = 0.01

It is possible that the ultimate populations, which have not yet been analysed, will be smaller than those of the control. Nevertheless, these results appear somewhat remote from those of Jatala et al. (15). The various P. lilacinus isolates may show different levels of aggressivity. Other species of Paecilomyces may be more effective, e.g. P. nostocoides, found by Dunn (7) on H. zeae cysts. Thorough research will be needed.

2.4 Can Resistant Varieties Be Selected?

Very few studies have been undertaken on the behaviour of different varieties. Greco and Lamberti (10) note several very slight differences in tolerance betwen certain varieties. Out of 15 varieties, Greco and Brandonisio (12) failed to find any which gave a satisfactory yield in the presence of H. carotae. All hybrids currently on the market are vulnerable.

Research could be done into resistant plant material. Biological know-how and technical expertise in the artificial inoculation of carrot seedlings have made a rapid test available.

Larvae are placed on the apex of the radicle of a seedling grown in a Petri dish, following the method devised by Mugniery and Person (18) for several Heteroderae. Certain precautions are necessary in the choice of larvae and the density of the inoculant: a proportion of the population is epigenetic, its sex being determined, inter alia, but such factors as competition, the age of the larvae and the plant itself (19).

After three weeks of growth at 20°C, the number of females formed per seedling is counted by observing the roots with a stereomicroscope without destroying the plant. It should thus be possible to select plants which inhibit the formation of females.

3. CONCLUSION

The biology and dynamics of H. carotae populations are such that the necessary crop rotation interval usually imposes excessive constraints on the farmer, who would prefer to apply a treatment, however onerous. The parasite cannot, however, be eradicated by chemical means, since the soil is recolonised. Fumigant treatments can protect the crop by sparing it from substantial attack by larvae while the plants are still young. Production of carrots for storage may result in a recolonisation giving higher populations at the end of the cultivation period than before treatment.

There is some hope for control by antagonists, but no technique applicable to agriculture is currently available; additional research will be needed to find strains or species of fungi sufficiently aggressive to provide protection even against slight infestations. In any event, the long term effects are totally unknown and should be studied.

Finally, there is one very interesting and promising line of research open into the control of H. carotae. Current levels of biological and technical expertise would appear to offer reasonable prospects of success in selecting plants resistant to the parasite.

REFERENCES

(1)   AMBROGIONI, L. and MARINARI PALMISANO, A., 1976. Effetto di avvicendamenti colturali su Heterodera carotae (Nematoda: Heteroderidae) e sulla produzione di carota in terreno infestato. Redia, 59: 355-368.

(2)   AUBERT, V., 1985. Hatching of Heterodera carotae. NATO advanced study institute on "cyst nematodes", Martina Franca, Italy. (in press)

(3)   BOSSIS, M., 1985. Quelques observations sur la dynamique des populations d'Heterodera carotae Jones (1950) dans l'ouest de la France. NATO advanced study institute on "cyst nematodes", Martina Franca, Italy. (in press)

(4)   CAUBEL, G., 1977. Les nématodes de la carotte. In La carotte: maladies et ennemis. INVUFLEC, PARIS.

(5)   CAYROL, J.-C. and FRANKOWSKI, J.-P., 1979. Une méthode de lutte biologique contre les nématodes à galles des racines appartenant au genre Meloidogyne. P.H.M. Revue Horticole, 193: 15-23.

(6)   CAYROL, J.-C. and FRANKOWSKI, J.-M., 1980. Connaissances nouvelles sur le champignon nématophage Artrobotrys irregularis (Royal 350). P.H.M. Revue Horticole, 203: 33-38.

(7)   DUNN, M.T., 1983. Paecilomyces nostocoides, a new hyphomycete isolated from cysts of Heterodera zeae. Mycologia, 75: 179-182.

(8)   GRECO, N. and LAMBERTI, F., 1977. Confronto di nematocidi nella lotta contro Heterodera carotae. Nematol. medit., 5: 1-9.

(9)   GRECO, N. and LAMBERTI, F., 1977. Ricerca delle dosi ottimali di alcuni nematocidi per la lotta contro Heterodera carotae. Nematol. medit., 5: 25-30.

(10) GRECO, N. and LAMBERTI, F., 1977. Suscettibilita di varieta di carota agli attacchi di Heterodera carotae. Nematol. medit., 5: 103-107.

(11) GRECO, N. and BRANDONISIO, A., 1980. Relationship between Heterodera carotae and carrot yield. Nematologica, 26: 497-500.

(12) GRECO, N. and BRANDONISIO, A., 1982. Comportamento di cultivar di carota in presenza di infestazioni di Heterodera carotae. Atti i congresso della societa italiana di nematologia. Redia, 65 "Appendice": 111-113.

(13) JATALA, P., KALTENBACH, R. and BOCANGEL, M., 1979. Biological control of Meloidogyne incognita acrita and Globodera pallida on potatoes. J. Nematol., 11: 303.

(14) JATALA, P., KALTENBACH, R., BOCANGEL, M., DEVAUX, A.J. and CAMPOS, R., 1980. Field application of Paecilomyces lilacinus for controlling Meloidogyne incognita on potatoes. J. Nematol., 12: 226-227.

(15) JATALA, P., SALAS, R., KALTENBACH, R. and BOCANGEL, M., 1981. Multiple application and long-term effect of Paecilomyces lilacinus in controlling Meloidogyne incognita under field conditions. J. Nematol., 13: 445.

(16) JONES, F.G.W., 1950. Observations on the beet eelworm and other cyst-forming species of Heterodera. Ann. appl. Biol., 37: 407-440.

(17) KERRY, B.R., 1984. Nematophagous fungi and the regulation of nematodes populations in soil. Helminth. Abstr., serie B 53: 1-14.

(18) MUGNIERY, D. and PERSON, F., 1976. Méthode d'élevage de quelques nématodes à kystes du genre Heterodera. Sciences Agronomiques Rennes, 217-220.

(19) MUGNIERY, D. and BOSSIS, M., 1986. Influence de l'hôte, de la compétition et de l'état physiologique des larves infestantes sur la pénétration, le développement et le sexe d'Heterodera carotae Jones. Nematologica (in press).

(20) OUDINET, R., CHERBLANC, G., SCHNEIDER, J. and DELOUSTAL, J., 1962. Quatre années d'essais de traitements contre le nématode de la carotte. Phytoma, 135: 11-15.

(21) OUDINET, R., 1968. Le nématode de la carotte. Phytoma, 198: 33-36.

(22) OUDINET, R., 1978. La carotte de Normandie dans le département de la Manche. Phytoma, 208: 3-12.

(23) WINSLOW, R.D., 1954. Provisional list of host plants of some root eelworms (Heterodera spp.) Ann. appl. Biol., 41: 591-605.

(24) WINSLOW, R.D., 1955. The hatching responses of some root eelworms of the genus Heterodera. Ann. appl. Biol., 43: 19-36.

Session 3
Tomato crops

Chairman: E.Brunel

Pest problems in field tomato crops in Spain

R.Albajes
Institut d'Investigació i Desenvolupament Agrari., Lleida Universitat Politècnica de Catalunya, Protecció de Conreus, Lleida, Spain
R.Gabarra, C.Castañe, E.Bordas & O.Alomar
Servei d'Investigació Agrària, Protecció de Conreus, Cabrils, Spain
A.Carnero
Instituto Canario de Investigación Agraria, La Laguna, Spain

Summary

Tomato is an important crop in Spain (60,000 ha in 1982), both outdoors and in greenhouses. The relative importance of pests is different in zones where outdoor crops coexist with greenhouse production than in other areas where no greenhouse cultivation exists. In the first ones, Trialeurodes vaporariorum, aphids (several species) and Liriomyza trifolii are the main pests. In the second ones, Tetranychus sp., Heliothis armigera and cutworms cause the main economic damage.
In Catalunya (first situation) a mirid bug Dicyphus tamaninii may be useful for decreasing greenhouse whitefly populations, but its feeding habits on tomato fruits may risk its possibilities as a biological control agent in tomato crops. It is necessary to implement a system for monitoring H. armigera and leaf-eating caterpillar populations in order to improve microbial insecticide sprayings.
Field tomato pests and their identified natural enemies in Canarias are similar to Catalunya ones, but the special climatology of the Islands (milder winter and summer) may modify the relationships described for Catalunya. Until now, research effort in Canarias has been directed to biological control in greenhouses with Encarsia formosa and Phytoseiulus persimilis.

1. Distribution of tomato cultivation in Spain

The tomato crop in Spain can be divided into three groups corresponding to different periods of harvesting (MAPA, 1984):
Group 1: Harvested between January 1 st. and May 31 st. In 1982 this covered a surface area of 8,698 ha.
Group 2: Harvested between June 1 st. and September 30 th. In 1982 this covered a surface area of 44,481 ha.
Group 3: Harvested between October 1 st. and December 31 st. In 1982 this covered a surface area of 7,423 ha.
Generally speaking, tomatoes in the first and third groups can be considered a semi-protected crop and those in the second group an outdoor crop. It should be noted, however, that in especially temperate areas the autumn harvest can be collected outdoors (particularly in the Canarias Islands), which would

Figure I. Main Spanish tomato cultivation regions.

increase the area of the tomato crop under cultivation in this
grouping.

From the point of view of pests and diseases, it is
important distinguish three kinds of areas, according to the
relative importance of the semi-protected and outdoor tomato
crop cultivations (Table I, figure I):

Area I: Here the main tomato crop is semi-protected, both
from the point of view of surface area as well as
gross production. Outdoor cultivation is relative
ly much less important. This is the case of the
province of Almeria with 5,000 ha of semi-protected crops and continous production from October to
May.

Area II: Almost complete outdoor cultivation, mainly
around the centre of Iberian Peninsula (Extremadu-
ra, Castilla-La Mancha, Aragón, Navarra-La Rioja
and Galicia). Most of the crop is produced for
processing.

Area III: Sequential cultivation in semi-protected and
outdoor conditions. Distributed along the
Mediterranean coast and in Canarias.

TABLE I. Tomato crop surfaces in outdoor and semi-protected
conditions in Spain in 1982 (MAPA, 1984)

REGION	KIND OF AREA	OUTDOOR SURFACE (ha)	SEMI-PROTECTED SURFACE (ha)
ALMERIA	I	800	5,000
GRANADA	III	1,350	70
MALAGA	III	1,202	1,060
MURCIA	III	1,783	3,039
CATALUNYA	III	3,360	221
VALENCIA	III	5,832	1,305
CANARIAS	III	3,519	460
EXTREMADURA	II	9,966	0
NAVARRA-LA RIOJA	II	1,707	270
ARAGON	II	2,778	0
CASTILLA-LA MANCHA	II	5,566	0
GALICIA	II	1,043	0

TABLE II. Main (XX) and secondary (X) pests of field tomato
crops in areas II and III (see text).

PEST	TOMATO CROPS OF AREA II	TOMATO CROPS OF AREA III
ACARINA, TETRANYCHIDAE Tetranychus sp.	XX	X
ACARINA, ERIOPHYIDAE Aculops lycopersici	X	X
HEMIPTERA, ALEYRODIDAE Trialeurodes vaporariorum	No problem	XX
HEMIPTERA, APHIDIDAE several species	X	XX
LEPIDOPTERA, NOCTUIDAE Heliothis armigera	XX	X
cutworms	XX	X
leaf-eating caterpillars	X	X
DIPTERA, AGROMYZIDAE Liriomyza bryoniae	No problem	X
L. trifolii	No problem	XX

2. Main pests in field tomato crops

The relative importance of pests in outdoor tomato crops
is strongly conditioned by their distribution in space and
time in relation to semi-protected crops.
As is well known, glasshouses provide pests with ideal
conditions for reproduction over the greater part of the year
-autumn, winter and spring. On the other hand, in our latitudes

199

summer in the glasshouses is too hot for pests from temperate climates to breed. In such conditions, a dual migratory movement of insects and mites is set up between the outside and inside of the glasshouses, which decisively conditions the population dynamics of the pests and consequently the control methods. This is the situation of the tomato crop in Area II. Outdoor cultivation generally covers the land in this area in a staggered fashion from the month of April until the month of October/November. It exists side by side with the semi-protected crop until the month of July and from October onwards. Moreover, in this area it is relatively common for outdoor glasshouses and plantations to be small and widely scattered, so that a mosaic of varying tomato crops exists throughout the whole year.

On the other hand, in Area II, located mainly in the centre of the Iberian Peninsula, pests are obliged to overwinter outdoors with the logical arrest in their development and high death rates, given the relatively low temperatures recorded in these regions in winter.

Table II include the main pests in the areas II and III, emphasizing their relative importance in each one of the two situations. The lack of published information in this respect must be noted. This has prevented us from carrying out a more detailed analysis, except for Catalunya and Canarias, for which more data is available. The data has been compiled from information received from colleagues in the different areas of cultivation.

It can be seen that most pests are common to both areas although their relative importance varies.

Tetranychids are more important in Area II fields where they must generally be treated with specific acaricides. The drier summer in the centre of the Peninsula probably favours the proliferation of these mites, compared with the higher relative humidities on the coast. It is also possible that the more intensive use of insecticides with an acaricide effect in Area III may have reduced the incidence of tetranychids in this latter area, although in some years specific acaricides have to be applied.

Aculops lycopersici (Massee) is widespread in most areas of Spain, although it is usually more harmful in the areas of València and Murcia.

Trialeurodes vaporariorum Westwood is an important tomato pest in Area III. It is this pest which probably best reflects the effect of the increase in protected land surface, given its high polyphagous nature. In this area it can be considered the pest requiring the highest amount of insecticide treatment and must constitute a priority target in research programmes on integrated control.

Different species of aphids colonize the outdoor tomato crops early in the year. Their incidence is very variable from one year to another and, generally speaking, they can be effectively controlled by a wide range of insecticides.

Among the lepidoptera we find three groups of noctuids which cause problems in field tomato crops in Spain- Heliothis armigera Hb., cutworms and leaf-eaters (mainly Autographa gamma (L.) and Chrysodeixis chalcites (Esp.)).

H. armigera has a variable importance, depending on the
year. Its economic incidence is greater in crops in Area II,
since it is generally well controlled by insecticides in those
in Area III.

Cutworms (mainly Agrotis segetum (Schiff.) and A. ipsilon
(Hfn.)) also present more serious problems in the Area II crops
than in those of Area III, especially in those reserved for
canning. In the latter, direct sowing favours the activity of
these soil noctuids. In transplanted crops, the soil is more
extensively tilled and in the time required for adults to lay
and the eggs to hatch, the plant has reached a stage of
development that makes less susceptible to the action of the
larvae. In addition, it is possible that the disinfection of
the soil, wich is often carried out in Area III crops, may lead
to a certain reduction in the non-migrant populations
(especially A. segetum).

The leaf-eating lepidoptera only occasionally manage to
cause significant damage and they are easily controlled by
insecticides.

Liriomyza bryoniae was practically the only agromyzid
species found in tomato crops of Area III some years ago. This
species only rarely caused significant damage in field crops.
In the last few years (earlier in the Canarias) the situation
has radically changed with the introduction of L. trifolii
Burgess. As expected with a recently introduced phytophagous,
L. trifolii has multiplied spectacularly along the Mediterranean coast, causing serious losses not only in the tomato crop
but in many others. Insecticide treatment has not proved able
controlling them. Although the problem apparently has decreased
in those areas w L. trifolii first penetrated, this new
species together with T. vaporariorum, represents today the
main problem to be considered in research programs on integrated control.

Some problems can occasionally be caused to tomato crops
by other phytophagous such as: Spodoptera littoralis (Boisd.)
(especially in the south of the Peninsula), Polyphagotarsonemus
latus (Banks), Acherontia atropos (L.) and several species of
trips.

It should be noted, finally, that pest problems in field
tomato crops in Spain derive from the ability of insecticide
products to control them. In Area II, the cost of treatment is
the limiting factor for pest control. In Area III, the problem
is the inability of insecticide products to control
T. vaporariorum and L. trifolii. This requires considerable
research effort in order to develop alternative methods to
chemical control, which will permit first an increase in crop
profitability and second a reduction of residues in tomatoes
for fresh consumption. The research work carried out in
Catalunya and Canarias, first aimed at semi-protected
cultivation, is currently directed to field tomato crops since
an interdependence of both biotopes seems clear. In fact, one
of the conclusions of the meeting of the Mediterranean Working
Group of IOBC on Protected Crops held at Catania in 1984
stressed the importance of "study of the epidemiology of the
pest populations and their natural enemies in field crops under
semi-protected conditions".

201

3. Present situation of research work in Catalunya

Catalunya, located in the north-east of the Peninsula, has three main areas of tomato production.

The first, and most important, of these is located around Barcelona. Most of the production is sent to the market of Barcelona and its metropolitan area (almost 4 million inhabitants). The Area III situation mentioned above is typically repeated here. The tomato plantations are small family plots with high productivity, distributed quite uniformly on the ground. Semiprotected crops (especially in spring) alternate with field crops (with staggered harvesting from   to October).

The second important area of tomato cultivation in Catalunya is the one surrounding Reus and Tarragona, with a similar situation to the above although with a smaller surface area under cultivation.

The third area is located in the Ebro delta. Cultivation was traditionally outdoor, in small family plots of tomatoes for fresh consumption and larger plots for canned tomatoes. In the last few years, early and late cultivation in glasshouses is increasing. This trend leads to a change in the pest situation from that typical of Area II to that of Area III.

According to a survey which we carried out among growers in these areas in 1983, greenhouse whitefly, different species of aphids (Macrosiphum euphorbiae (Thomas), Myzus persicae (Sulzer), Aphis sp.), and H. armigera were the main crop pests in the first two areas with an average of 15 insecticide treatments during cultivation. In the third area, cutworms, Tetranychus sp., H. armigera and leaf-eating lepidoptera were the pests with highest economic impact. The average number of insecticide treatments in this area was five.

However in 1985, L. trifolii seemed to be the most important pest in all three areas, particullarly in the third one.

Some tomato experimental plots with several spraying programs have been established at Cabrils Experimental Station (northern of Barcelona) since 1983 in order to determine the status of pests and their natural enemies under a low and no insecticide treatment regime.

The most significant results are summarised below (unpublished).

The lowest greenhouse whitefly population levels were observed in non-sprayed plots and, consequently, no damaged fruits with sooty mould were observed in these plots. A similar picture was noticed in commercial tomato and other vegetable fields which had not been sprayed with insecticides. However, when some insecticides were sprayed greenhouse whitefly population quickly increased and, later, fruits and leaves became damaged by sooty mould. In the present year, more than 70 % of tomato fruits have been damaged by sooty mould in spite of nine insecticide (methomyl and pyrethroids) treatments.

Experimental plots which had not been treated with insecticides showed relatively high population levels of a mirid bug, Dicyphus tamaninii Wagner. Its population trends were significantly correlated with whitefly population. When the mirid bug was excluded from tomato plants by muslin cages, greenhouse whitefly population increased to damaging density, but if the mirid was introduced inside the cages, the whitefly decreased drastically in a few weeks. Furthermore, D. tamaninii has been observed predating on eggs, nymphs and adults of T. vaporariorum in laboratory experiments.

However, D. tamaninii has also phytophagous habits. It feeds on tomato fruits which are depreciated for fresh consumption. Discoloured spots surrounded by an exceptionally red halo appear in the feeding punctures. In experimental plots where mirid bug populations were high, a noticeably quantity of damaged fruits appeared. In laboratory experiments it could be observed that the mirid was able to complete a generation on tomato fruits.

In all experimental plots, aphids (mostly M. euphorbiae, M. persicae and A. fabae) colonized tomato plants early in the season. Pirimicarb sprayings controlled aphid populations, but when no insecticides were applied they also decreased abruptly. In all cases a low number of mummies and specific predators were recorded.

During 1983, 84 and 85, visual counting of eggs on plants and light and pheromone (Zoecon(R) and INRA) trapping of adults were used in order to monitor the tomato worm populations. However, no significant relationship could be established among the three methods after the three years, since population numbers were extremely low. On the contrary, a high number of eggs and larvae of H. armigera were positively assessed on carnation buds and flowers during the same period. Probably, carnation is preferred to tomato by the pest and the abundance of that crop in the region could explain the low population numbers in tomato plots.

A relatively high density of leaf-eating larvae (A. gamma and/or C. chalcites) was only recorded in 1983. Then, two sprayings with Bacillus thuringiensis Berliner (Thuricide (R)) prevented economic losses. The effect of these treatments on H. armigera could not be evaluated due to the low level of this pest. No effect of B. thuringiensis on non lepidopterous species was detected. In 1984 and 1985 a noticeable quantity of leaf-eaters were parasited by Apanteles sp. and Litomastix sp. or they were affected by diseases (mostly caused by virus).

4. Preliminary outlook for integrated control in tomato field crops in Catalunya

T. vaporariorum and recently L. trifolii are the main pest species that should be taken into account for integrated control programs in field tomato crops in Catalunya.

Adults of the first species leave greenhouse early crops and colonize field crops during the last part of June and all July, so a sudden increase of outdoors populations may be recorded at this time. In these conditions whitefly control in the field should ideally be based on two different tactics: first, to reduce the populations of whitefly in early protected crops, and second, to increase the number of the pest's natural enemies in the field.

A certain number of predators of whitefly belonging to different families have been found in Catalunya, especially in crops little treated with insecticides (Bordas et al. 1985). Among the parasitoids Encarsia tricolor Förster is the most frequent one.

In accordance with the results summarized above, the mirid bug D. tamaninii is an efficient predator of whitefly in non treated plots. Other mirids have been reported as aleyrodid predators (Khristova et al. 1975; Tsybulskaya and Kryzhanovskaya, 1980). D. tamaninii is a polyphagous predator. In addition to predation on whitefly, it has been observed predating on noctuid eggs, aphids, mites and leafminer larvae in the field and laboratory.

Unfortunately it is well known that many species of mirids share predatory and phytophagous habits, even at the same developmental stage, so their use in biological control may prove to be limited (Carayon, 1971; Fauvel, 1983). Some mirid species, taxonomically very close to D. tamaninii, have been quoted as vegetable pests in the mediterranean area: Macrolophus nubilus Herrich-Schaeffer in Spain (Gómez-Menor, 1954) and Cyrtopeltis tenuis Reuter in Egypt (El-Dessouki et al 1976), both in tomato.

However, no significant correlation between D. tamaninii population levels and number of damaged fruits may be found in experimental plots. It seems that arthropod preys are preferred to tomato fruits by D. tamaninii and it is only when preys drop below a threshold density that they cease to be preferred to tomato fruits. The above results and the fact that D. tamaninii may be found in non treated commercial plots (in greenhouses and in outdoors) and the fact that damage is seldom detected in fruits, may encourage future experimental work about its ecology and feeding behaviour.

Recent introduction of L. trifolii in Catalunya may enhance the possibilities of the mirid as a biological control agent in field vegetable crops. In fact, Parrella et al.(1982) pointed to Cyrtopeltis modestus (Dist.), a taxonomically close mirid, as a possible biological control agent of L. trifolii in California.

On the contrary, Encarsia formosa Gahan has shown fewer possibilities for whitefly control in field conditions. In the last years some experimental releases have been done, but they failed to control whitefly populations. Now this species is being introduced in early protected crops, so that migration from greenhouses to field might be reduced.

E. tricolor is a spontaneous aleyrodid parasitoid in mediterranean area and it may be also used successfully in protected crops (Albajes et al. 1980). Its high fecundity and development rate, studied in laboratory conditions, are encouraging (Artigues and Avilla, personal communication), but some aspects of its behaviour and sex determination should be known in order to improve breeding techniques and to determine its usefulness for greenhouse whitefly biological control. The use of E. tricolor in spring protected crops would make possible to colonize outdoors with a higher parasitoid populations.

In experimental plots, aphids showed significant

TABLE III. Main common and specific pests of vegetable crops
in Canarias

		TOMATO	WATER-MELON	MELON	PUMPKIN	PEPPER	EGG-PLANT	FRENCH BEANS	CUCUMBER
APHIDIDAE	Aphis gossypi	x	x	x		x			
	Macrosiphum euphorbiae	x	x	x	x				
	Myzus persicae	x	x	x	x	x	x	x	x
	Aulacorthum solani	x							
	Aphis fabae	x					x	x	
ALEYRODIDAE	Trialeurodes vaporariorum	x	x	x	x	x	x	x	x
NOCTUIDAE	Heliothis armigera	x							
	Autographa gamma	x							
	Chrysodeixis chalcites	x							
AGROMYZIDAE	Liriomyza trifolii	x	x	x					
ERIOPHYIDAE	Aculops lycopersici	x							
TETRANYCHIDAE	Tetranychus urticae	x	x	x	x	x	x	x	x
TARSONEMIDAE	Polyphagotarsonemus latus			x					
OTHERS	(Lizards, birds)	x							

population levels only in May and June. After, they practically
disappear. Only a few mummies and no specific predators were
observed. The mentioned survey among growers indicated that
aphid populations were present in tomato crops during August
and September and several insecticide sprayings were necessary
for controlling them. The observation of D. tamaninii predating
on aphids in laboratory and some observations in the field,
allow us to assume that this mirid could be the main factor in
decreasing aphid populations. The early presence of aphids in
tomato plots may attract first mirid adults from surrounding
spontaneous plants. In fact, several workers quote the role of
mirids on aphid populations (Dirimanov and Dimitrov, 1975;
Mc Gavin, 1982).

In order to implement an IPM program for tomato crops it
is necessary to use selective insecticides for lepidopterous
pests (H. armigera and leaf-eating caterpillars). On that
respect B. thuringiensis and especially viral insecticides may
prove their usefulness. Since only young larvae are susceptible
enough to field sprayings, a pest population monitoring system
should be implemented.

The presence, in Spanish Mediterranean coast, of Diglyphus
isaea (Wlk.) and Dacnusa sibirica Telenga, two important
parasitoids of L. trifolii, may contribute to solve the
problems caused by this leafminer, but more research effort
will be necessary in the years to come on this area.

TABLE IV. Identified parasites and predators of vegetable
pests in Canarias.

PEST	PARASITES	PREDATORS
Trialeurodes vaporariorum	Encarsia formosa E. partenopea E. tricolor Encarsia sp. (Hym; Aphelinidae)	Chrysopa sp. (Neur; Chrysopidae)
Aphids	Aphidius matricariae Diaretiella rapae Praon volucre (Hym; Aphidiidae) Entomophthora planchoniana E. aphidis (Entomophthoraceae)	Aphidoletes aphidimyza (Dip; Cecidomyiidae)
Tetranychus urticae		Iphiseius degenerans Amblyseius sp. (Acarina, Phytoseiidae)
Liriomyza trifolii	Diglyphus isaea Diglyphus sp. (Hym; Eulophidae)	

5. Pests and their natural enemies in Canarias

The climatology of Canarias offers optimum conditions for
continuous development of pest populations of tomato and other
vegetable and ornamental crops. The mean temperatures of the
coldest (January) and hotest (August) months are 16.8ºC and
23.6ºC respectively. The mean relative humidity is 70 %.

Table III shows the main pests of the most important
vegetable crops. Several parasites and predators of these pests
have been identified (Table IV). In general they are the same
found in Catalunya, but the special climatology of Canarias may
modify their interrelationships.

Most research efforts in Canarias have been devoted to
biological control of greenhouse pests. E. formosa (spontaneous
species in Canarias) and Phytoseiulus persimilis Athias-Henriot
have been introduced in experimental tomato and rose
greenhouses with encouraging results (Carnero and Barroso,1985).
Since L. trifolii was introduced in Canarias several years ago,
some experimental works about its ecology have been carried
out.

REFERENCES

ALBAJES, R. et al. (1980). La mosca blanca de los invernaderos Trialeurodes vaporariorum en El Maresme II. Utilización de Encarsia tricolor (Hym.; Aphelinidae) en un invernadero de tomate temprano. An.INIA Ser.Agr.16: 135-145.

BORDAS, E. et al. (1985). La lutte integrée dans les cultures maraîchères en Catalogne: présent et futur. Bull. OILB/SROP VIII/I:1-9.

CARAYON, J. (1961). Quelques remarques sur les hemiptères- heteroptères, leur importance comme insectes auxiliaires et les possibilités de leur utilisation dans la lutte biologique. Entomophaga 6: 133-141.

CARNERO, A. and BARROSO, J.J. (1985). Control biológico de Trialeurodes vaporariorum (Hom.; Aleyrodidae) en las Islas Canarias. Bolm.Soc.port.Ent. 1:323-331.

DIRIMANOV, P. and DIMITROV, L. (1975). Role of useful insect in the control of Thrips tabaci Lind. and Myzodes persicae Sulz on tobacco. VIII Int.Pt.Prot. Congress. Moscow. vol II:71-72.

EL-DESSOUKI, S.A., EL-KIFL, A.H. and HELAL, H.A. (1976). Life cycle, host plants and symptoms of damage of the tomato bug Nesidiocoris tenuis Reut. (Hemiptera: Miridae) in Egypt. Zeit. Pflazenkrant.Pflazensch. 83:204-220.

FAUVEL, G. (1983). Des punaises utiles, tiens donc! in "Faune et flore auxiliaires en agriculture" pp.71-77. ACTA. Paris s.d. 368 pp.

GOMEZ-MENOR, J. (1954). Un mírido que ataca al tomate y al tabaco. Bol.Pat.Veg. y Ent.Agr. 21: 193-200.

KHRISTOVA, E., LOGINOVA, E., PETRAKIEVA, S. (1975). Macrolophus costalis Fieb., predator of whitefly (Trialeurodes vaporariorum Westw) in greenhouses. VIII Int.Pl.Prot.Congress. Moscow. vol. II: 124-125.

M.A.P.A. (1984). Anuario de Estadística Agraria. Ministerio de Agricultura, Pesca y Alimentación. Madrid. s.f. 682 pp.

Mc Gavin, G.C. (1982). A new genus of Miridae (Hem.; Heteroptera). Entomologist's Monthly Magazine 118 (1412/1415):79-86.

PARRELLA, M.P. et al. (1982). Control of Liriomyza trifolii with biological agents and insect growth regulators. California Agriculture 36(11/12): 17-19.

TSYBULSKAYA, G.N. and Khryzhanovskaya, T.V. (1980). A promising insect control agent. Zashchita Rastenii. nº10,23 (in russian).

Development of a control programme against the aphids *(Macrosiphum euphorbiae* Thomas and *Myzus persicae* Sulzer) in processing tomato cultures in the south-east of France

R.Bues, J.F.Toubon & H.S.Poitout
INRA, CRA d' Avignon, Station de Zoologie et d' Apidologie, Montfavet, France

Summary

In the south of France, three main pests of processing tomato can be observed on the aerial parts of the plant: the aphids, Macrosiphum euphorbiae and Myzus persicae, the tomato fruitworm, Heliothis armigera, and the mites, Vasates lycopersici and Tetranychus turkestani. Yet, only aphids are regularly treated. They are present on the plant from May to June. The sampling method used is based on the observation of 1 leaflet of the terminal leaf of one branch. A correlation is observed (r = 0.928 N = 212) between the number of aphids per leaflet and the mean number of leaflets with apterous aphids. Experiments on sub-plots resulted in provisional treatment threshold at 5-10 aphids per leaflet. A sequential sampling plan (based on presence/absence) was developed and tested in 1985 and compared with a simple sampling technique based on the arbitrary selection of leaflets.

INTRODUCTION

Between 5 000 and 7 000 hectares are planted with processing tomatoes each year in France. The annual crop of 300-400 000 tonnes, for the most part produced in three regions south of the 45th parallel (South-East, Roussillon and South-West), is processed by approximately 50 factories. In 1984, 75.28% of the national crop (355 817 tonnes) came from the South-East (15).

Fungicides and insecticides are applied repeatedly during the entire growing season, from May to September/October. There are three main types of animal pest: insects in the soil (nematodes, wireworms and terricolous fruitworm), insects attacking the stalk and foliage (aphids, mites) and those attacking the fruit (tomato fruitworm). As a generalisation, nevertheless, three major pests can be identified: aphids (Myzus persicae Sulzer and Macrosiphum euphorbiae Thomas), the tomato fruitworm (Heliothis armigera Hbn.), though its economic significance varies from year to year and from place to place, and mites (Vasates lycopersici Mass., Tetranychus turkestani (Ugar. and Nik.)).

The population dynamics of these three pests have been monitored on a regular basis since 1981. More detailed study of sampling methods and treatment thresholds has been carried out for two of these, aphids and fruitworms. This paper is devoted exclusively to the results concerning aphids.

There are two types of aphid damage to tomatoes (9):

- direct damage, in which the foliage and particularly the leading shoots shrivel up, the flowers fail to develop and the plant's growth is stunted;
- indirect damage, with the deposition of honey-dew and fumagine and, especially, the transmission of viruses. Studies carried out jointly with plant pathologists have shown no direct relationship between the size of populations and the percentage of plants showing symptoms of necrosis. Systematic treatment against aphids, in order to avoid contamination, is expensive and ineffective (19, 11). Moreover, tomatoes are more susceptible to aphid damage at the plantlet stage than when they are putting out the first floral clusters (8), the stage at which transplanting is usually done.

The two aphid species, M. persicae and M. euphorbiae, are often observed together on tomato crops (13, 5). Both have been extensively studied, particularly in relation to vegetable crops: on potatoes (3, 14, 12), on tomatoes (4, 21, 22) and on cabbages (17, 18).

The results of the work described here underpin the development of a control programme, at the plot level, based on a simplified sampling method.

METHODOLOGY

a)   Experimental Plots

Studies were carried out on two experimental plots. One, Domaine des Vignères, is an approximately one-hectare plot situated 20 km from Avignon and run as a commercial venture. The crop density is 40 000 plants/hectare. Spray irrigation is provided and pesticides are administered at a rate of 200 litres/hectare using tractor-mounted spray booms.

Work on this plot from 1982 to 1985 involved studying population dynamics, sampling and the introduction of a control programme. The second plot, Domaine Saint Paul (10 km from Avignon) covers an area of 1 000 m^2. In order to make observation easier, the density of this plot is 20 000 plants/hectare. The crop is watered by gravity or trickle irrigation and spraying is done using a back-pack system at a rate of 1 000 litres/hectare. This plot is divided into 16 sub-plots, four each of the following four variants: three levels of the populations being studied (these being defined in the text), and an untreated control. Quantification of differences between the variants was achieved by comparing the weight of marketable tomatoes produced by each plant. Trials on this plot were carried out from 1983 to 1985.

The results should make it possible to identify the population level at which the yield is affected (= treatment threshold).

In the two trial plots spraying, for cryptogamic and bacterial diseases, is carried out with Captafol and copper-based products respectively; aphids are attacked using pirimicarb (37.5 g active constituent/hl).

b)   Sampling Method

From preliminary observations it was known not only that aphids were spread uniformly over all the leaflets forming the tomato leaf, but also that there was a preferential concentration of apterous aphids on the terminal leaves of the plant. As soon as the plants were sufficiently mature, however, sampling trials were carried out at two strata: at the upper stratum, the terminal leaflet was picked from the last properly formed leaf at the end of a branch; at the lower stratum, the terminal leaflet was taken from a leaf in the midst of the foliage.

To take each sample, the leaflet was inserted into a polyethylene tube (diameter 3 cm, length 7 cm) and broken off against the edge of the tube, which was then immediately stoppered. Counting was done using a binocular magnifying glass. The two species of apterous aphids were counted separately but the data were merged in the interests of drawing up the control programme. In the Vignères experimental plot, the sampling unit at each stratum comprised five samples taken over a distance of one metre. Each week, 14 sampling units (70 leaflets) were studied. The row numbers were drawn by lot to determine the sampling locations and in each row a number of paces was selected such that 3-4 points were sampled in each row.

Earlier studies at the Saint Paul plot had shown that in view of the small area of the plots, a satisfactory enumeration of the population in each variant could be achieved by taking 28 leaflets at each stratum, i.e. 7 leaflets in each sub-plot.

Parallel to the sampling of wingless and winged aphid populations in the plots, a Moericke yellow trap covered in yellow paint (Ripolin 514) was installed on the edge of the plots. The captured insects were collected twice a week.

RESULTS AND DISCUSSION

A)   Population Dynamics of Winged and Wingless Aphids

Figure 1 shows the population dynamics for apterous aphids in non-treated plots at Saint Paul from 1983 to 1985 and the curves for winged aphids caught in the yellow trap at the same site.

The presence of apterous aphids was recorded from the end of May to the end of June, the maximum being during the first 20 days of June. In 1983 and 1984, almost the entire apterous population comprised M. euphorbiae; in 1985, approximately 20% were M. persicae. Predation and parasitism, although more evident in June - especially in 1985 - cannot fully explain the abrupt fall in populations at the end of June. This coincided with an increase in average temperatures to about 20°C, implying maximum temperatures of 25°C or more. It has been shown that these two species cannot develop at temperatures above 25-30°C (1). Other factors may, however, be involved, such as the physiological state of the plant. Similar observations have been made with regard to other aphid species on strawberries (19).

A study of the capture curves for the trap shows that flights increased at the end of May as temperatures climbed above an average threshold of approximately 15°C. The maximum incidence of M. euphorbiae flight in 1984, during the second fortnight in May, coincided with a period of heavy rainfall (121 mm from 11 to 30 May) and relatively low average temperatures, which could explain the small number of apterous aphids observed during this period. It should be noted that during this same year, M. persicae were captured three times as frequently as M. euphorbiae, even though the former were virtually absent from the crop.

The conclusion of significance to a control programme is the regular presence of apterous aphids on processing tomato crops from the end of May to the end of June. Less detailed work carried out in 1981-1982 confirms this fact.

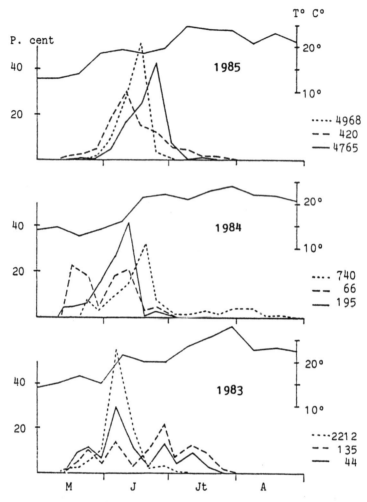

Fig. 1. Comparison of Distribution Curves for Apterous Aphids Observed on
Untreated Control Plots (...) and Captures of M. euphorbiae (---)
and M. persicae (___) in the Moericke Trap (Saint Paul)

B)   Sampling of Apterous Aphid Populations

a)   Comparison of two sampling levels

        Figure 2 shows the results of the 1984 comparison between the number
of apterous aphids observed at each of the two sampling strata (upper and
lower) in plots receiving differing numbers of sprayings. The same variation
in population was found irrespective of the sampling level. As a general
rule, however, the population is proportionately greater in upper-stratum
samples, which is, incidentally, the location of the most significant direct
damage. It is therefore possible to make a representative assessment of
apterous aphid populations by sampling this stratum alone.

No. apterous aphids
(Log)

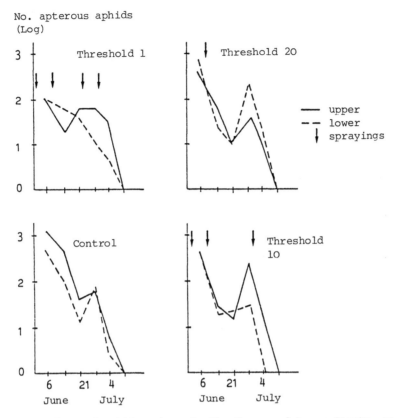

Fig. 2. Comparison of Aphid Numbers in the Upper and Lower Strata, on
Untreated Control Plots and on Plots Managed at 3 Treatment
Thresholds: 1, 10 and 20 Aphids per Leaflet (Saint Paul, 1983)

b)   Presence/absence sampling

Figure 3, showing the correlation between the logarithms for average
values and for variance in the number of apterous aphids sampled from 1982
to 1985, confirms the aggregative nature $(\sigma^2 > \bar{x})$ of aphids. In terms of a
control programme, it is unrealistic to plan on counting the aphids every
week in order to establish the pre-spraying population level. Over the trial
period, it was evident that a correlation existed between the number of
aphids and the percentage of leaflets found to harbour aphids. Figure 4
shows this linear regression (winged and wingless aphids) for 212 samples
taken between 1982 and 1985. At 0.928, the correlation coefficient is
extremely significant.

This correlation demonstrates that simple presence/absence sampling
can provide an approximate enumeration of the population on the crop. It
should be pointed out that the assessment is least accurate with very low or
very high populations, i.e. at the two ends of the regression line. It is at
these two extremes of aphid-affected leaflets, however, that it is easiest
to decide whether or not the crop needs spraying. This is also apparent from
the observed quadratic correlation (Figure 5) between the percentage of

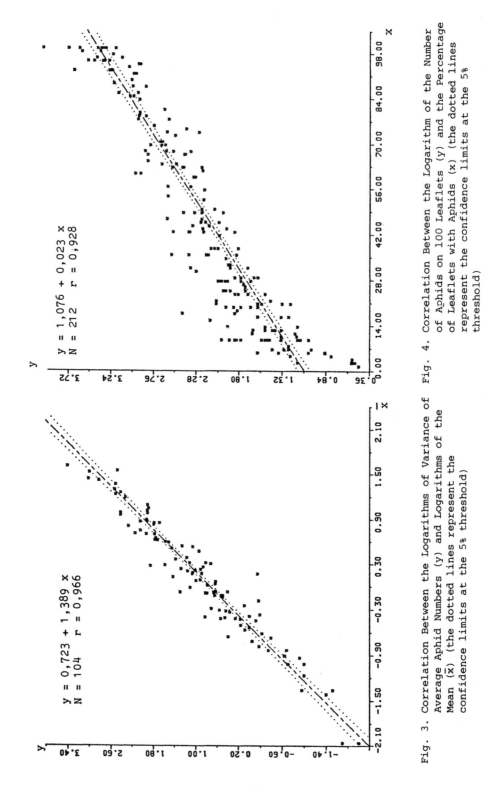

Fig. 3. Correlation Between the Logarithms of Variance of
Average Aphid Numbers (y) and Logarithms of the
Mean ($\bar{x}$) (the dotted lines represent the
confidence limits at the 5% threshold)

Fig. 4. Correlation Between the Logarithm of the Number
of Aphids on 100 Leaflets (y) and the Percentage
of Leaflets with Aphids (x) (the dotted lines
represent the confidence limits at the 5%
threshold)

214

leaflets with aphids and the logarithm of the variance in the number of leaflets with aphids. The variance is lowest (less than 1) where the percentage of leaflets with aphids is either very low or very high. The highest variance (2.70) occurs where 50% of leaflets harbour aphids. In this case, it is necessary to examine approximately 11 sampling units of leaflets to achieve a 5% risk and a precision of ±1 leaflet. However, this result requires verification in large plots.

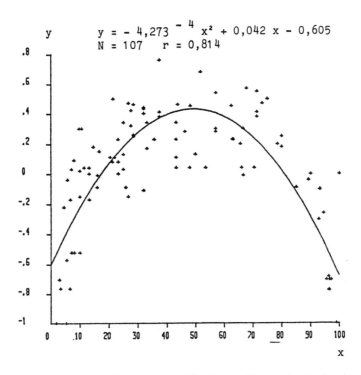

$$y = -4,273^{-4} x^2 + 0,042 x - 0,605$$
$$N = 107 \quad r = 0,814$$

Fig. 5. Quadratic Correlation Between the Logarithms of the Variance of the Number of Leaflets with Aphids (y) and of the Percentages of Leaflets with Aphids (x)

c) Presence/absence sequential sampling

   Initiation of a control programme for aphids on tomatoes requires prior sampling of the crop. The time required for this sampling is one of the factors limiting the institution of a control programme. Sequential sampling, perfected for a very large number of insects, makes it possible almost to halve the number of samples. This method has been used for sampling M. persicae on sugar-beet (16), Brevicoryne brassicae L. and M. persicae on Brussels sprouts (20) and M. euphorbiae on potatoes (21).
   Two sequential sampling methods have been used: one based on the assumption that a negative binomial distribution exists, and requiring counting of the aphids on each sample; the other based on the assumption of a binomial distribution, this requiring only determination of the presence or absence of aphids (16). Another sequential sampling method, already

215

proposed (6, 7) and described (2), does not require adjustment on the basis of a theoretical frequency distribution. This latter method, which we used and tested in 1985, determines the spatial distribution of an organism by using the average crowding index: $\overset{*}{x}$. This aggregation index is calculated as follows:

$$\overset{*}{x} = \overset{*}{x} + \frac{(\sigma^2 - 1)}{\bar{x}}$$

where $\bar{x}$ equals the average density and σ^2 the variance.

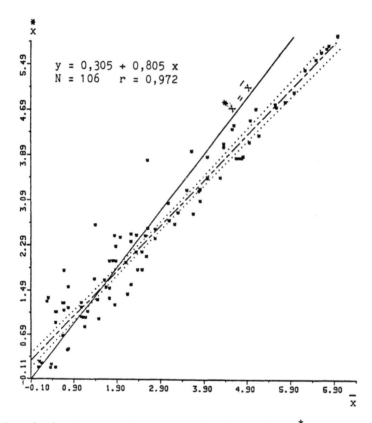

Fig. 6. Correlation Between the Average Crowding Index ($\overset{*}{x}$) and the Average Number of Leaflets Harbouring Aphids per Sampling Unit of 5 or 7 Leaflets ($\bar{x}$) (the dotted lines represent the confidence limits at the 5% threshold)

Figure 6 shows the linear relationship between this aggregation index, calculated on the basis of the average variance in the number of leaflets infested with aphids, and the average number of leaflets with aphids in each sampling unit of 5 or 7 leaflets. It corresponds to a linear model (r = 0.972) where the slope (0.805) represents the distribution coefficient and the ordinate (0.305) a contagion index. Where the slope is 1 and the ordinate 0 ($\bar{x} = \overset{*}{x}$), the distribution of insects can be regarded as random

(6). Figure 6 shows that there is a progression from a very dispersed distribution, corresponding to primary contamination with few infestations, to the distribution over the whole of the plot typical of widespread attack by secondary contamination. The transition from one distribution pattern to the other occurs at an average of 2 aphid-infested leaflets per sampling unit. In order to have more data available, however, these results were merged with those for sampling units of 5 or 7 leaflets. Although this makes the threshold for widespread aphid attack less precise, a generalised dispersion through the plot can be assumed when between 60 and 80% of leaflets harbour aphids, corresponding to 3-8 aphids per leaflet.

Moreover, trials have been carried out since 1983 with the aim of determining a treatment threshold. Although these ongoing studies have yet to identify a precise threshold, it is nevertheless now possible to state that a population of between 5 and 10 aphids per leaflet can be tolerated by tomato crops without any significant reduction of the yield.

These thresholds correspond (see Figure 4) to approximately 70.5 and 73.7% respectively of leaflets with aphids, amounting to 3.5 and 4 affected leaflets per sampling unit of 5 leaflets, this being quite close to the above theoretical figures.

These treatment thresholds allow the calculation of upper and lower acceptability limits according to the formula

$$C = N \times TT \pm t \sqrt{N((\alpha + 1) TT + (\beta - 1) TT^2)}$$

where C = the number of leaflets with aphids, N = the number of sampling units studied, TT = treatment threshold, t = the value of the t-test (1.96) at the 5% limit for an infinite number of degrees of freedom, and α and β respectively the ordinate and the regression slope, $\bar{x} - \overset{*}{x}$ (Figure 6). The curves for the two levels are shown in Figure 7. One drawback of sequential sampling is that where the average of the population studied is the same as the economic threshold, a large number of leaflets have to be examined if one is to emerge from the zone of indecision between these two curves. It is possible to calculate, for a given confidence interval, the maximum number of sampling units observed (N max) at which the population level becomes the same as the economic threshold:

$$N (max) = \frac{t^2}{d^2} ((\alpha + 1) TT + (\beta - 1) TT^2)$$

where d is the accepted confidence interval.

In this case, at a confidence interval of 0.5 leaflets per sampling unit, 33 and 37 sampling units will be required for the 5 and 10 aphid-per-leaflet thresholds respectively. In 1985 at the Vignères site, 44 sequential samples were taken at either the 5 or 10 threshold; 38 were in agreement with conventional sampling carried out at the same time, the decision to spray was twice taken when the threshold number of apterous aphids had not yet been reached and four such decisions were taken after the maximum number of samples.

CONCLUSIONS

The results obtained in 1982-1985 enabled the development of a simple sampling method based on observing the presence or absence of aphids during a specific part of the growing season for processing tomatoes (from the end of May to the end of June). At present, the sequential sampling method is based on a provisional treatment threshold of 5 or 10 aphids per leaflet, i.e. 70.5 and 73.5% of leaflets harbouring aphids. Trials carried out in 1985 on commercial crops showed that a threshold of 10 aphids per leaflet

217

did not affect crop yields. Nevertheless, further trials are necessary to verify this result and to examine the influence on the treatment threshold of the plant variety and its phenologic stage at the time of aphid attack.

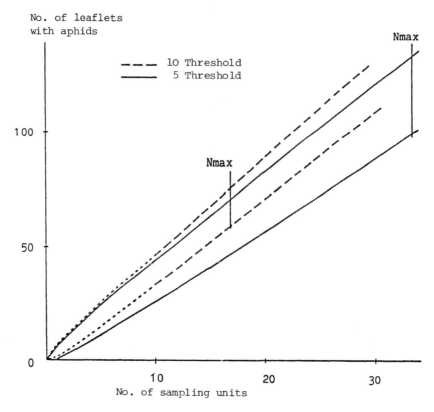

Fig. 7. Upper and Lower Limits of Acceptability for the Thresholds of 5 and 10 Aphids on Average per Leaflet, i.e. 70.5 and 73.5% of Leaflets Harbouring Aphids (sampling unit = 5 leaflets)

ACKNOWLEDGEMENTS

The authors wish to thank Mr. F. Leclant and Mr. M. Renoust from the research laboratory of the zoology faculty at Montpellier (INRA) for their indispensable assistance in the trials, as well as Mrs. Jacquemond and Mr. J.P. Leroux from the phytopathology laboratory at Avignon (INRA) for their help in identifying the plants suffering from virus attacks.

REFERENCES

(1)   BARLOW, C.A., 1962. The influence of temperature on the growth of experimental populations of Myzus persicae (Sulzer) and Macrosiphum euphorbiae (Thomas) (Aphididae). Can. J. Zool., 40, 145-156.

(2)   BOIVIN, G. and VINCENT, C., 1983. L'échantillonnage séquentiel en phytoprotection, revue de la méthode. Agric. Can, 29, 2, 114-126.

(3) ELLIOT, W.M., 1973. A method of predicting short term populations trends of the Green peach aphid, _Myzus persicae_ (Homoptera : aphididae) on potatoes. Can. Entomol., 105, 11-12.

(4) ELLIOT, W.M., 1981. The relationship of embryo counts and suction trap catches to populations dynamics of _Macrosiphum euphorbiae_ (Homoptera : aphididae) on tomatoes in Ontario. Can. Entomol., 113, 1113-1122.

(5) FLINT, M.L., (Ed.), 1985. Integrated pest management for tomatoes. Univ. California, Dw. Agric. Nat. Resour. (Publ. 3274), Oakland, 104 p.

(6) IWAO, S., 1968. A new regression method for analyzing the aggregation pattern of animal populations. Res. Popul. Ecol., 10, 1-20.

(7) IWAO, S., 1975. A new method of sequential sampling to classify populations relative to a critical density. Res. Popul. Ecol., 16, 281-288.

(8) JACQUEMOND, M., 1982. L'ARN satellite du virus de la mosaïque du concombre. IV. Transmission expérimentale de la maladie nécrotique de la tomate par pucerons. Agronomie, 2, 7, 641-646.

(9) LECLANT, F., 1982. Les effets nuisibles des pucerons sur les cultures. 37-56. _In_ : A.C.T.A., Les pucerons des cultures. Journées d'Etudes et d'informations, Paris, 2-4 March 1981. A.C.T.A., Paris, 351 p.

(10) LLOYD, M., 1967. Mean Crowding. J. Anim. Ecol., 36, 1-30.

(11) MARCHOUX, G., LECLANT, F. and LECOQ H., 1984. Rôle des Aphides dans l'épidémiologie des maladies à virus des cultures maraîchères. Bull. Soc. entomol. Fr., 89, 716-730.

(12) PETIT, F.L. and SMILOWITZ, Z., 1982. Green peach Aphid feeding damage to potato in various plant growth stages. J. econ. Entomol., 75, 431-435.

(13) RABASSE, J.M., LAFONT, J.P. and MOLINARI, J., 1985. Colonisation et développement des populations de pucerons sur tomate en serre dans le sud de la France. Bull. OILB/SROP, 8, 1, 27-42.

(14) ROBERT, Y., 1980. Recherches sur les pucerons en Bretagne. Thèse Doct. ès. Sci. nat., Univ. Rennes I, 242 p.

(15) SONITO (Société Nationale Interprofessionnelle de la Tomate), 83, route de Lyon, 84000 Avignon. Documents internes.

(16) SYLVESTER, E.S. and COX, E.L., 1961. Sequential plans for sampling Aphids on Sugar Beets in Kern County California. J. econ. Entomol., 54, 1080-1085.

(17) TRUMBLE, J.T., 1982(a). Within-plant distribution and sampling of Aphids (Homoptera : Aphididae) on Broccoli in Southern California. J. econ. Entomol., 75, 587-592.

(18) TRUMBLE, J.T., 1982(b). Aphid (Homoptera : Aphididae) population. Dynamics on Broccoli in an Interior Valley of California. J. econ. Entomol., 75, 841-847.

(19) TRUMBLE, J.T., OATMAN, E.R. and VOTH, V., 1983. Thresholds and sampling for aphid in strawberries. Calif. Agric., 37, 11/12, 20-21.

(20) WILSON, L.T., PICKEL, C., MOUNT, R.C. and ZALOUM, F.G., 1983. Presence-Absence sequential sampling for Cabbage Aphid and Green peach Aphid (Homoptera : Aphididae) on Brussels sprout. J. econ. Entomol., 76, 476-479.

(21) WALKER, G.P., MADDEN, L.V. and SIMONET, D.E., 1984(a). Spatial dispersion and sequential sampling of the potato aphid. _Macrosiphum euphorbiae_ (Homoptera : Aphididae), on processing-tomatoes in Ohio. Can. Entomol., 116, 1069-1075.

(22) WALKER, G.P., LOWELL, R.N. and SIMONET, D.E., 1984(b). Natural mortality factors acting on potato aphid (_Macrosiphum euphorbiae_) populations in processing-tomato fields in Ohio. Environ. Entomol., 13, 724-732.

The present status on bacterial diseases of tomatoes in Greece

C.G.Panagopoulos
Laboratory of Plant Pathology, Athens College of Agricultural Sciences, Athens, Greece

So far the following bacterial diseases of tomatoes have been
diagnosed in Greece: Bacterial Canker caused by Corynebacterium michiganense pv. michiganense (first recorded in 1957), Bacterial Speck caused by Pseudomonas syringae pv. tomato (first recorded in 1971), Bacterial wilts caused by Pseudomonas solanacearum (first recorded in
1970) and Erwinia chrysanthemi (first described in 1982) (2), Bacterial Spot caused by Xanthomonas campestris pv. vesicatoria (first recorded in 1976) and Pith Necrosis or Brown Pith Necrosis (first recorded in 1977). Some studies of their symptomatology, etiology and
control have been carried out. Stem lesions are a very common symptom
of Bacterial Speck infections and may be easily confused with other
tomato diseases (especially late blight and Virus Streak diseases).
Therefore, accurate diagnosis of each particular disease is necessary
in order to take as soon as possible the appropriate control measures.

A recent study in Greece on the etiology of pith necrosis revealed that the causal agent is either Pseudomonas viridiflava or another
pseudomonad of the syringae group, and that the existence of pathogenic isolates of Pseudomonas fluorescens should be investigated (1).

Bacterial Canker and Speck diseases are the most important, economically, mainly because they cause considerable crop losses and
are very common throughout the Country. Control is based on the use
of disease-free seed or seed treatment, destruction of crop residues,
use of proper sanitary measures and copper compounds and/or antibiotics sprays (against Speck disease).

The following commercial cultivars and hybrids were evaluated
for their level of resistance to artificial inoculation of C. michi-

ganense pv. michiganense and P. syringae pv. tomato: Roma VF , ACE 55 VF , Macedonia, Heinz 2274, Super California, ES 58, BOG-AT 69, Early Pack, A 200, 44024 VF and Ponterossa. The cultivars Macedonia and ACE 55 FV, seem to be resistant to Speck disease, the Heinz 2274 to Canker disease, while the cultivar BOG - AT 69 was found resistant to both diseases.

REFERENCES

1. ALIVIZATOS, A.S. (1984). Aetiology of tomato pith necrosis in Greece. Proc. 2nd Working Group on Pseudomonas syringae pathovars, pp 55-57, Sounion, Greece.

2. ALIVIZATOS, A.S. (1985). Bacterial wilt of tomato caused by Erwinia chrysanthemi. Plant Pathology (in press).

Disease resistance: An excellent resource for world-wide biological control in tomato

M.Cirulli & F.Ciccarese
Dipartimento di Patologia Vegetale, Università di Bari, Italy

1. Summary

Disease resistance is commonly considered a highly effective and satisfactory means of control because it is inexpensive, easy to apply, and not harmful to other biological components of our ecosystem. It is used on a world-wide scale in cultivated herbaceous and arboreous species to control the most destructive diseases.

Disease resistance has been found in many cultivated plants such as vegetables, grains, oil plants, fruits, fiber plants, stone fruits, small fruits, forage plants, sugar plants and many others.

The amount of information on fungal, bacterial, viral, nematode and non- parasitic disease resistance in tomato is great and highly specific. We have attempted to illustrate disease resistance in tomato in a general way, paying particular attention to the research areas with which we are most familiar, i.e. Fusarium wilt and mosaic-TMV.

For some diseases, such as those caused by Leivellula taurica (Lév.) Arn., Septoria lycopersici Speg. and Sclerotium rolfsii Sacc. only preliminary information is available and no commercial varieties have yet been obtained. However, many other diseases, such as wilts, mosaic-TMV and leaf mould, are now fully understood and modern cultivars possessing resistance genes against these diseases have been developed since the 1950s.

The use of resistant cultivars constitutes the main factor for yield increase in tomato. For instance, in the 1940s tomato yields ranged between 10-20 tons per hectare on the average. Later, in the 1950s, and 1960s, the finding of the I-1 and Ve genes for resistance to Fusarium and Verticillium wilts and the consequent release of cultivars with combined resistance to these diseases enabled growers to achieve yields of over 50-60 tons per hectare.

Regarding tomato, in the past 50 years great progress has been made in the study of: the mechanism of pathogen variability, the distribution and identification of physiological races of pathogens, and the use of techniques for screening and evaluating germplasm collected in many parts of the world and in the center of origin of this species. In the light of the knowledge derived from these studies, it was possible to carry out extensive research on the variability of numerous populations of wild and cultivated Lycopersicon species. As a result, a large number of genes for resistance to fungi, virus, bacteria, nematodes and physiological disorders

Table I - Oligogenic and polygenic resistance found in cultivated and wild species of *Lycopersicon*.

Diseases or pathogens	L. esculentum	L. pimpinellifolium	L. peruvianum	L. hirsutum	L. glandulosum	L. chilense	L. hirsutum var. glabratum
Fusarium wilt	Polygenic	I-1, (I-2)	-	-	-	-	-
Verticillium wilt	Polygenic;Ve	-	-	-	-	-	-
Pyrenochaeta lycopersici	Polygenic	-	+	-	+; (pyl)	(pyl)	+
Cladosporium fulvum	Cf_n	Cf_n	Cf_n	Cf_n	Cf_n	-	Cf_n
Phytophthora infestans	-	Ph-1,Ph-2;Polygenic	-	+	-	-	-
Oidiopsis taurica	-	+	+	+	-	-	-
Stemphylium spp.	-	+; Sm	+	-	-	+	+
Alternaria spp.	Monogenic;Polygenic	-	-	-	-	-	-
Rhizoctonia solani	Monogenic;Polygenic	-	-	-	-	-	-
Colletotrichum spp.	+	+	-	-	-	-	-
Phoma destructiva	+	-	-	-	-	-	-
Sclerotium rolfsii	+	-	-	-	-	-	-
Septoria lycopersici	-	-	(Se)	(Se)	-	-	-
Botrytis cinerea	+	-	-	-	-	-	-
Pseudomonas solanacearum	Polygenic	-	-	-	-	-	-
P. syringae pv tomato	Pto	Pto	+	-	+	-	+
Xanthomonas campestris pv tomato	+	-	-	-	-	-	-
Corynebacterium michiganense	Polygenic	Polygenic	-	Polygenic	-	-	-
Tobacco mosaic virus (TMV)	-	(Tm-1) (Tm-2) Tm-2^a	(Tm-1) (Tm-2) Tm-2^a	Polygenic (Tm-1) (Tm-2)	-	-	-
Tomato yellow leaf curl virus	-	Monogenic	-	-	-	-	-
Spotted wilt virus (SWV)	-	+	-	-	-	-	-
Tomato etch virus (TEV)	+	+	-	+	-	-	-
Meloidogyne spp.	-	+	(Mi)	+	+	-	+
Blotchy ripening	+	-	-	-	-	-	-
Fruit pox	+	-	-	-	-	-	-
Fruit cracking	+	-	-	-	-	-	-

+ = Resistance reported, but not definitively studied.

- = Resistance not described yet.

( ) = Species of probable origin of known resistance.

are now available to plant breeders for the release of multiple disease resistant cultivars that contribute to minimizing losses, reducing the cost of production, and favouring expansion of tomato production areas (Table 1).

2. Resistance to tomato pathogens

2.1. Fusarium wilt

Fusarium wilt of tomato (<u>Fusarium oxysporum</u> f. sp. <u>lycopersici</u> Sn. et Hn.) is present all over the world. It is a vascular disease that causes leaf yellowing and wilting, stem discoloration and death of the plants (Figure 1): the same symptoms occur in the Verticillium wilt from which it is not easily distinguishable. The control of Fusarium and Verticillium wilts can be obtained either by soil disinfestation or by the use of resistant tomatoes.

A polygenic resistance against Fusarium wilt was found by Pritchard in

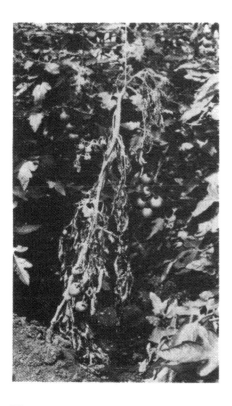

Figure 1 – Fusarium wilt.

Table II - Types of resistance to Fusarium
wilt in tomato.

- OLIGOGENIC

Resistance genes	Source and reference
I-1	- *Lycopersicon pimpinellifolium* (Bohn e Tucker, 1940)
I-2	- *L. pimpinellifolium x L. esculentum* (Stall & Walter, 1965; Cirulli & Alexander, 1966)

- POLYGENIC

| | - *Lycopersicon esculentum* (Pritchard, 1922) |

1927, and transferred into such varieties as Marglobe and Rutgers (Pritchard, 1927) which are still available today (Table II).

The first oligogenic resistance (monogenic dominant I-1 gene) was derived from a selection of Lycopersicon pimpinellifolium (Bohn and Tucher, 1940) and is widely used in tomato-breeding programs. More than 20 years later, after the appearance of a new race of the fungus (Stall, 1961) virulent against I-1 gene resistant varieties, a new oligogenic resistance monogenic dominant was identified (Stall, 1961; Stall and Walter, 1965), designed as I-2 (Cirulli and Alexander, 1966) which is highly effective against race 1 (sensu Gabe, 1975) of. F. oxysporum f. sp. lycopersici.

Race 1 was reported in the U.S.A. (Stall, 1961; Goode, 1966; Miller and Kananen, 1968), Israel (Katan and Wahl, 1966), Brazil (Tokeshi et al., 1966; Kurozawa and Pavan, 1982), Morocco (Laterrot and Pecaut, 1966), Republic of South Africa (Holz, 1976), France (Bouhot et al., 1978) and Senegal (Collingwood and Defranco, 1979). Preliminary studies seemed to indicate a third race of the fungus in Brazil (Tokeshi et al., 1966) which was also isolated in Western Florida, U.S.A., and Queensland, Australia from commercial fields of tomato varieties possessing the I-2 gene (Volin and Jones, 1982; Grattidge and O'Brien, 1982). According to the international code for race designation, this race should be indicated as race 2 (Table III). A search for tomatoes resistant to the third biotype of Fusarium has been undertaken and the accessions P.I. 143524, and P.I.

221838 produced mixed responsed when tested to the Florida race 2 isolates (Volin and Jones, 1982). In a screening test done in Australia, the tomato lines US-638 proved to possess field resistance to the third race and good commercial traits (Grattidge, 1982).

Table III - Host-parasite relationship in the Fusarium wilt of tomato

- OLIGOGENIC RESISTANCE

Variety (genes for resistance)		Parasite (races)*		
		0 (1)	1 (2)	2 (3)
Bonny Best Marmande		S	S	S
Pan American Ohio WR	I-1	R	S	R
Walter	I-1	R	R	S

- POLYGENIC RESISTANCE

Variety

"Marglobe"
"Rutgers"
"Pritchard"
"Break O'Day"

R = Resistant; S = Susceptible
* = Upper row: race classification *sensu* Gabe (1975)
     Lower row: old race classification

2.2. Verticillium wilt

Verticillium wilt of tomato is widespread in the temperate countries of the world and is caused by the two species Verticillium albooatrum Rke et Berth., more adapted to the humid and colder climate of Northern Europe, and Verticillium dahliae Kleb., more common in warmer areas.

The first Verticillium wilt-resistant tomato, Reverside, was released by Lesley and Shapovalov (1937) and Shapovalov and Lesley (1940). The resistance of this tomato was probably polygenic. In 1951, Schaible et al., found a monogenic dominant type of resistance which was designated with the Ve symbol. The source of this resistance was a small-fruited tomato type called wild Perù, Utah selection 665. The Ve gene is commonly incorporated into most commercial tomatoes.

A physiological race (denominated race 2) of Verticillium pathogenic against Ve-resistant tomatoes was found in the USA (Alexander, 1962; Bender and Shoemaker, 1977; Hall and Kimble, 1972), Europe (Cirulli, 1969; Tjamos, 1981) and Australia (O'Brien and Hutton, 1981).

A field resistance to race 2 of V. dahliae, presumably of the polygenic type, was described by Hubbeling, Alexander and Cirulli (1971) for the "Heinz 1350" tomato; this, however, was only partial and depended to some extent on the soil type and pH. In a screening of L. esculentum cultivars and breeding lines, line MEL-2668170 G was selected as the best source of resistance (Okie and Gardner, 1982). Resistance in MEL appeared to be recessive and governed by 3 or less genes operating against North Caroline isolates of V. dahliae race 2. Using a Brazilian isolate of race 2, Laterrot (1984) found that MEL-2668170 G was susceptible and that a line of L. esculentum, ITRAT-L3, possessed a good level of resistance which appeared to be monogenic and dominant.

2.3. Pyrenochaeta lycopersici Schneider et Gerlach

The corky root disease (Figure 2) caused by Pyrenochaeta lycopersici is usually present in the greenhouse (Cirulli and Ciccarese, 1983) and was recently observed to affect field-grown tomatoes, as well (Campbell et al., 1982).

Several sources of resistance to corky root are reported in tomato. On crossing the highly resistant Lycopersici hirsutum with the tomato rootstock KNVF, a F1 hybrid was obtained with resistance to P. lycopersici and also to Verticillium, Fusarium and Dydimella lycopersici. Moderate resistance to corky root was found in L. glandulosum and L. chilense, in which it was monogenic and recessive, being controlled by the same gene pyl (Laterrot, 1983). In a search for new sources of resistance against TMV, Leivellula taurica and P. lycopersici, Cirulli (1968) found that several P.I. of L. peruvianum, G. glandulosum and L. hirsutum var. glabratum possess high levels of resistance to corky root (Figure 3). Some of these accessions exhibited plants that were resistant to both TMV and P. lycopersici.

Volin and Mc Millan (1978) reported that L. esculentum accessions P.I. 260397, P.I. 262906 and P.I. 203231 showed resistant reactions in the field and that heritability in the broad sense was 25-43% with a minimum of 4-8 genes influencing resistance.

Laterrot (1972) found a moderate resistance in the tomato "Espalier", which was controlled by a major partially dominant gene.

In Italy, Ciccarese and Cirulli (1983) found that the resistance of the two breeding lines "Bari 80201" and "Bari 80101 derived from L. hirsutum var. glabratum (Ebben et al., 1978) and L. glandulosum (Hogenboom, 1970) respectively and attributed to single, partially dominant genes was very promising. A higher level of resistance was observed in L. esculentum breeding lines and F$_1$ hybrids derived from the latter wild species of tomato (Ciccarese and Cirulli, 1983).

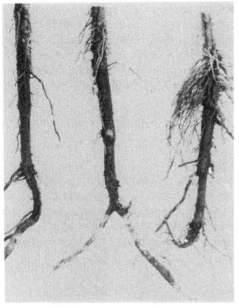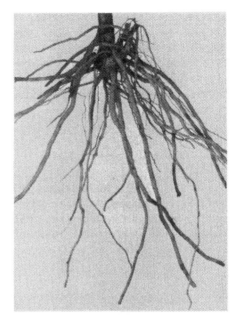

Fig. 2 – Susceptible reaction to corky root in *Lycopersicon esculentum*.

Fig. 3 – Resistance reaction to corky root in *Lycopersicon peruvianum*.

2.4. Rhizoctonia solani Kuhn.

Fruit rot disease under particular climatic conditions, especially high humidity, may cause considerable damage.

Barksdale (1974) described that segregants of an F_2 population from Merit (susceptible) x P.I. 193407 (resistant line) of L. esculentum consisted of plants whose reaction to the disease ranged over the whole parental disease range, thus indicating that resistance was probably polygenic. Werner et al. (1980) working on tomato lines of L. esculentum found that the tolerance in U.S.A. 75B 846-1-1 was controlled by a single recessive gene and that the tolerance in USDA 75B 610-3 was polygenic.

2.5. Phytophthora infestans (Mont.) de Bary

Under favorable cool temperatures and heavy rainfall, P. infestans causes severe epidemics. Considerable efforts have been made to control this disease by breeding for resistance. Two types of resistance to late blight fungus have been described. A dominant-gene resistance, Ph-1 (= R_1t, TR_1), was obtained from a cross of L. esculentum x L. pimpinellifolium, P.I. 204996 (Bonde and Murphy, 1952; Graham, 1955; Gallegly and Marvel, 1955; Gallegly, 1964). This resistance is effective against race 0 of

P. infestans. After the discovery of a physiological race (race 1) pathogenic to tomatoes carrying the Ph-1 gene (Gallegly, 1952; Graham, 1955), a new search for resistance resulted in the finding of a polygenic resistance in P.I.204994 which confers partial resistance to both races 0 and 1.Turkensteen (1973) found that the tomato line W.Va 700, which was derived from P.I. 204994, possessed partial resistance coltrolled by a single gene designated as Ph-2.

2.6. Didimella lycopersici Kleb.

The disease affects stems, fruits and less frequently the leaves. Stem lesions are the most common symptom and may lead to plant wilt. Bouxema (1982) found resistance to Dydimella stem rot in L. hirsutum and L. hirsutum var. glabratum. Genetic studies involving these two wild species and L. esculentum showed that resistance is dominant and non-monogenetic.

2.7. Cladosporium fulvum Cooke.

The leaf mould disease is common in greenhouses all over the world. Attacks are favored by high humidity (80-100%) and temperatures of 18-24°C, which frequently occur in greenhouses.

Resistant cultivars have been the most effective and widely-used means of tomato leaf mould control. Many resistance genes designated as Cfn (n = progressive numbers assigned to Cf genes matching new physiologic races of C. fulvum) have been found. Commonly, a few years after the introduction of resistant cultivars of tomato, new physiologic races of the pathogen occurred (Table IV).

Table IV - Determination of physiologic races of tomato pathogens exemplified through the gene-for-gene relationship of the Cf resistance factor and the pathotypes of Cladosporium fulvum.

Dominant resistance genes in tomato differentials	Physiologic races							
	0	1	2	3	1.2	1.3	2.3	1.2.3
cf	S	S	S	S	S	S	S	S
Cf_1	R	S	R	R	S	S	R	S
Cf_2	R	R	S	R	S	R	S	S
Cf_3	R	R	R	S	R	S	S	S
$Cf_1 Cf_2$	R	R	R	R	S	R	R	S
$Cf_1 Cf_3$	R	R	R	R	R	S	R	S
$Cf_2 Cf_3$	R	R	R	R	R	R	S	S
$Cf_1 Cf_2 Cf_3$	R	R	R	R	R	R	R	S

S = Susceptible;   R = Resistant.

The first resistance factors to the tomato leaf mould organism were found in the L. esculentum cultivar Sterling Castle, gene Cf_1 (Alexander, 1934), and in L. pimpinellifolium, known at the time as L. racemigerum: Sengbush and Laschakova-Hasenbush, 1932) Cf_2. Great efforts to control C.fulvum by breeding for resistance have been made in Ontario, Canada (Bayley, 1947), resulting in the release of cultivars possessing single genes for resistance, e.g. Vetomold (Cf_2), P-136 (Cf_4), V-121 (Cf_3), and more than one gene for resistance, e.g. Vinequeen (Cf_3, Cf_4), V-548 (Cf_1, Cf_2, Cf_4), Vantage (Cf_2, Cf_4). Vagabond (Cf_1, Cf_2), Vantage (Cf_1, Cf_4). Resistant single-gene cultivars quickly succumbed to new races of the pathogen whereas those with more than one gene for resistance were resistant for a longer period of time (Kerr and Bayley, 1964).

Resistance Cf genes have been found in many wild specie of tomato, e.g. L. pimpinellifolium (cvs Improved Bay State, V-545), L. hirsutum (V-501) and L. peruvianum (cvs V-542, Vantage, Waltham, Mold Forcing No. 22).

Figure 4 – Susceptible (left) and resistance (right) reactions to *Cladosporium fulvum*.

According to Pearson (1959), in a system based on specific gene-for--gene interactions, for 5 Cf genes of resistance to tomato leaf mould the occurrence of 32 (= 2) physiological races of C. fulvum can be expected. Up to present time, 25 Cf genes have been found (Kanwar et at.,1980 and 1980a). The naming of races has been done in two ways: in Canada they have been numbered in the chronological order of appearance (Patrick et al., 1971), whereas in Europe have been named according to the corresponding gene (or genes) for resistance. The latter have been divided into five groups A,

B, C, D and E: group A includes "old races" (0, 1, 2, 3, 1.2, 1.3, 2.3 and 1.2.3) and the other groups include "new races" (4, 1.4, 3.4, etc.) (Hubbeling, 1968).

2.8. Stemphylium spp.

The gray leaf spot can cause serious defoliation of tomatoes, particularly when the humidity is high. This disease is caused by Stemphilium solani Weber (= S. floridanum Hannon et Weber) and S. botryosum f. sp. lycopersici Rotem, Cohen and Whal. The symptoms caused by these species are rather similar and detailed observation is necessary to determine the inciting organism.

In genetic studies involving a resistant line of L. pimpinellifolium and a susceptible cultivar of L. esculentum, Hendrix J.W and Frazier W.A. (1949) demonstrated that the resistance is governed by a single dominant gene to which the symbol Sm was assigned.

Tomato cultivars possessing the Sm gene for resistance to S. solani were also found to have a high degree of resistance to S. floridanum Hannon and Weber and S. botryosum f. sp. lycopersici (Bashi et al., 1973). Data obtained from F$_1$ hybrids between resistant ("Anahu", "Walter", and "Tropic") and susceptible ("Eilion" and "Campbell 1327") tomatoes indicated that resistance is dominant. Individual plants of several accessions of L. hirsutum f. glabratum, L. peruvianum, L. chinense and L. pimpinellifolium also showed higher levels of resistance than L. esculentum cultivars against S. floridanum and S. botryosum f. sp. lycopersici.

2.9. Alternaria spp.

Alternaria solani (Ell. et Mart.) Jones et Grout

This pathogen mainly causes two distinct disease syndromes: a leaf spot and defoliation phase, which may be found at any stage of plant growth but more frequently as the leaf tissues reach maturity, and a stem canker phase, usually referred to as collar rot, which may occur on young seedlings in plant beds.

Some of the early work on resistance to early blight has been reported by Moore and Reynard (1945), Andrus et al. (1942), Reynard and Andrus (1945). Reynard and Andrus (1945) reported that resistance to the collar rot phase of A. solani was conferred by a single recessive gene. Genetic studies done on the L. esculentum tomato lines 71-B2, CI 943 and Anahu-R indicated that 71-B2 contains 2 or more recessive factors for resistance to the leaf spotting or defoliating phases of early blight (Barksdale and Stoner, 1977). Barksdale and Stoner (1973) described a horizontal resistance which conferred a high level of field tolerance to A. solani on the foliage and near immunity to stem canker. Factors for resistance to these 2 phases of the disease are independently segregated from one another. In a glasshouse test of 30 cultivars, Vakalonnakis (Greece, 1983) found 28 cvs very susceptible to A. solani while "Meltine" and "Nemato" were slightly less susceptible (70-72% disease compared with 88-100%).

Isolates of A. solani vary in their pathogenic characteristics. Bonde (1929) reported the existence of physiological races and Henning and Alexander (1952) described at least seven races of A. solani based on pathogenicity tests.

Alternaria alternata (Fr.) Keissler (= A. tenuis Auct.) f. sp. lycopersici.

This fungus is a distinct pathotype capable of primary infection of leaves, stems and fruits. Typically, the disease causes dark-brown to black cankers on the stem near the soil line or on the stem above the ground. Gorgan, Kimble and Misaghi (1975) found that only 25% of 265 freshmarket cvs tested were susceptible, and attributed the resistance to a single dominant factor. Among resistant cvs were Ace, Florida MH-1, Floradel, Manapal, Walters, 6718 VF-F$_1$, Marvel and Marglobe. These authors also found that of the 40 processing cultivars screened, 28 were resistant and 12 susceptible.

2.10. Septoria lycopersici Speg.

This disease mainly causes leaf spotting and defoliation especially where temperatures are mild and rain abundant (Figure 5).

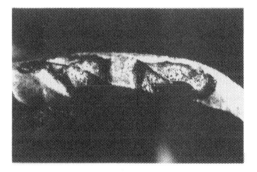

Figure 5 – Attack of *Septoria lycopersici* on tomato leaf.

In 1945 Reynard and Andrus screened 127 cultivars of L. esculentum for resistance to Septoria and found them to be highly susceptible. However, some resistance was found in 12 of 267 accessions whose ancestry included L. hirsutum and L. peruvianum. Resistance studies done by the above mentioned authors, in crosses involving resistant accessions and commercial cultivars, showed that resistance (represented by restricted type of leaf lesion) was conferred by a single dominant gene to which the symbol Se was assigned.

Barksdale and Stoner (1978) reported a high level or resistance to S. lycopersici in the L. esculentum tomato PI 422397. The resistance reaction consisted of necrotic flecks on the leaves on which few, if any, pycnidia developed; where they did form sporulation was poor. This resistance was inherited as a single dominant gene.

2.11. Pseudomonas solanacearum (E.F. Smith) E.F. Smith

This bacterium is common in tropical climates and cause the bacterial wilt disease of tomato. Lin et al. (1974) found that of 522 varieties tested, 4% were highly resistant. The L. esculentum cultivars UP-11670, Palmar, Nematex, UC 8-1-21, Venus and UPCA 20-29 were the most resistant ones. Graham and Yap (1976) in diallel cross study of 6 L. esculentum cultivars (Walter, CRA-66, H-7741, Venus, VC-4 and Llanos de Colce) representing a range of susceptibility to resistance (99.5 to 20.8 on a resistant scale) found that general combining ability was considerably more important than specific combining ability and concluded that inheritance of resistance was mainly attributable to additive gene action. Sonoda et al. (1980) reported that under the environmental conditions of Florida, the highest resistance occurred in H 7997, CRA 66 and PI 126408 tomatoes.

2.12. Pseudomonas syringae pv tomato (Okabe, 1933) Young, Dye and Wilkie 1978.

Bacterial speck occurs in temperate and subtropical areas of the world and is favored by high humidity and rainfall. The disease consists of small, brown to black necrotic spots on the leaves, petioles, peduncles, stems and fruits (Figure 6).

In Ont. 7710, an experimental cultivar of L. esculentum with complex ancestry, Pitblado and McNeill (1983) found a single dominant gene, Pto, which governs resistance (immunity under field conditions) to the bacterial speck pathogen P. syringae pv tomato. High resistance (symptomless reaction) was also described by Pilowsky and Zutra (1982) in accessions of L. pimpinellifolium, L. peruvianum, and L.hirsutum var. glabratum. These authors stated that L. esculentum cultivars Rehovot-13 and Hosen-Eilon, previously reported as having resistance, proved to be susceptible to isolate 134 of the bacterium.

Pilowsky and Zutra (1982a) also suggested that a different single dominant gene confers resistance to Ontario 7710 and in L. pimpinellifolium P.I. 126430.

In a study concerning the inheritance of resistance in L. pimpinellifolium P.I. 126937 and Ontario 7710, Laterrot (1983a) found that resistance in P.I. 126937 is governed by a single dominant allel, which is at the same locus as Pto.

Figure 6 – Leaf spots caused by *Pseudomonas syringae* pv *tomato*.

2.13. Corynebacterium michiganense subsp. michiganense (Smith 1910) Jensen 1934.

This bacterium causes local infections, consisting of necrotic spotting on leaves and fruit; and systemic infection producing vessel browning, wilting and plant death (Figure 7).

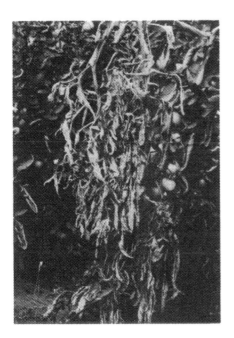

Figure 7– Bacterial canker of tomato caused by *Corynebacterium michiganense*. Left: wilt of a plant caused by systemic infection. Right: canker and production of adventitious root primordia in an infected stem.

Thyr (1972) found resistance to bacterial canker in L. pimpinellifolium and L. hirsutum. From tests with isolates differing in virulence, he obtained evidence that resistance was of the horizontal type. Based on his genetic studies Thyr (1976) also concluded that three different sources of resistance, namely Bulgaria-12 of L. esculentum, P.I. 330727, and Utah 737 (P.I. 344103) and Utah 20 (P.I. 344102) of L. pimpinellifolium, possessed an incompletely dominant polygenic type of resistance.

De Jong and Honma (1976) stated that the resistance of Bulgaria 12 could be explained by a 4-gene model: one recessive, and three dominant. They also found resistance to bacterial canker in L. hirsutum that could be explained by the same 4-gene model requiring an additional se/se allele for resistance.

2.14. Tomato mosaic virus (TMV)

The Tobacco Mosaic Virus (TMV) is widespread throughout the world causing high crop losses both in the greenhouse and in the filed. This virus is well known for its stability in high temperatures and aging, remaining active in plant residues (Brodbent, 1964) and soil (Doolitle, 1928). It is transmitted mechanically by epidemics spreading rapidly from a few diseased plants. The disease causes mosaic (Figure 8), flower abortion and high reduction of yield. At present, the most satisfactory disease control has been afforded by the use of resistant cultivars.

So far, TMV resistance has been described several time in tomato (Table V). Frazier and Dennet (1949) made an interspecific cross between Lycopersicon hirsutum, Lycopersicon peruvianum and Lycopersicon esculentum. Working on tomato lines derived from Frazier and Dennet's material two single genes, Tm-1 and Tm-2, were found by Holmes (1954) and Soost (1963), respectively.

Alexander (1963) found a high type of resistance in Lycopersicon peruvianum, P.I. 128650. This re-

Figure 8 - Symptoms of mosaic (TMV) in a tomato leaf.

236

sistance was studied by Cirulli and Alexander (1969) and attributed to a single gene indicated as Tm-2^a.

Table V - TMV resistance genes in tomato

Resistance genes	Source and reference	
Tm-1	Complex cross between L. hirsutum L. peruvianum L. esculentum Frazier Dennett (1949)	(Holmes,1952)
Tm-2	Frazier & Dennett's (1949) complex cross	(Soost, 1963)
Tm-2^a	L. esculentum x L. peruvianum	(Cirulli & Alexander, 1969)

Isolates of TMV possessing diversified virulence have been observed in many parts of the world. Cirulli and Ciccarese (1975) reported that best strain classification of TMV is obtained at 26°C using differential tomatoes with resistance genes in the homozygous condition (Table VI).

Table VI - Criteria proposed for classification of strains of TMV
(from Cirulli and Ciccarese, 1975)

Resistance gene in the differential hosts	TMV strain nomenclature (a)							
	0	1	2	2^a	1.2	1.2^a	2.2^a	1.2.2^a
Susceptible	S	S	S	S	S	S	S	S
Tm-1/Tm-1	-	S	-	-	S	S	-	S
Tm-2/Tm-2	-	-	S	-	S	-	S	S
Tm-2^a/Tm-2^a	-	-	-	S	-	S	S	S
Tm-1/Tm-1 Tm-2/Tm-2	-	-	-	-	S	-	-	S
Tm-1/Tm-1 Tm-2^a/Tm-2^a	-	-	-	-	-	S	-	S
Tm-2/Tm-2 Tm-2^a/Tm-2^a	-	-	-	-	-	-	S	S
Tm-1/Tm-1 Tm-2/Tm-2 Tm-2^a/Tm-2^a	-	-	-	-	-	-	-	S

(a) S = Susceptible reaction
   - = Resistance reaction

2.15. Tomato yellow leaf curl virus (TYLCV)

This virus causes leaf curling and yellowing, and flower sterility. Pilowsky and Cohen (1973, 1974) found resistance to this disease in L. pimpinellifolium, LA-121. Inheritance studies in crosses derived from this species and the susceptible "Pearson" of L. esculentum showed, by inoculating the test plants with the vector tobacco whitefly Bemisia tabaci Gennadius, that the resistance to TYLCV was incompletely dominant and monogenic.

2.16. Meloidogyne spp.

The root knot of tomato is caused by the following species of Meloidogyne: M. incognita (Kofoid et White) Chitwood, M. incognita-acrita Chitwood, M. javanica (Trenb.) Chitwood, M. arenaria (Neil) Chitwood, and M. apla Chitwood. Typical symptoms occur on roots which enlarge into galls after infection and frequently cause the roots to have a clubbed appearance (Fig. 9).

Figure 9 – Root knot caysed by *Meloidogyne* (right).
Resistance reaction (gene M̲i̲) in the left.

Bailey (1940), Romshe (1942), Ellis (1943), Mc Farlane et al. (1946) and Watts (1947) supplied the first indication that resistance could be found in L. peruvianum. Lines of L. peruvianum vary in their resistance to root knot. Taylor and Chitwood (1951) described a selection of L. peruvianum which was resistant to M. incognita but moderately susceptible to M. arenaria and M. incognita acrita. Sasser (1954) observed that L.

238

peruvianum may be susceptible to M. incognita-acrita and M. javanica but highly resistant to M. incognita. Thomason and Smith (1957) found lines of L. peruvianum that were highly resistant to M. javanica and M.incognita-acrita.Taylor et al. (1955), in a cooperative screening test involving 144 tomato accessions including L. hirsutum, L. hirsutum var. glabratum, L.peruvianum, L. glandulosum and natural crosses of L. esculentum x L. pimpinellifolium, found resistance to Meloidogyne in L. peruvianum and, to a lesser extent, in L. glandulosum. Using embryoculture, Smith (1944) obtained a fertile F_1 hybrid between the L. esculentum cultivar Michigan State Forcing x L. peruvianum (P.I. 128657). Genetic studies carried out on several tomato lines, all derived from this original cross, showed that resistance is controlled by a single, dominant gene (Gilbert and McGuire, 1956; Barham and Winstead, 1957; Winstead and Barham, 1957; Thomason and Smith, 1957) to which the symbol Mi was given (Gilbert and McGuire, 1956). This gene is effective against M. incognita, M. incognita-acrita, M. javanica and M. arenaria, but not M. hapla.

In greenhouse tests, Riggs and Winstead (1959) showed that repeated passages of the same population of Meloidogyne may give rise to new pathotypes. Although populations of M. incognita occurred in nature (Winstead and Riggs, 1963), no physiological race of Meloidogyne has assumed great economic relevance in nature.

3. Factors affecting the phenotypic expression of resistance in tomato

Environmental conditions have long been known to exert a great influence on plant disease. Already at the time of Theophrastus (370-286 BC), cereal crops growing in elevated lands and subjected to wind were considered less liable to rust   than those grown at a lower altitude. In the eighteenth century, Tillet (1955) reported that the causes of wheat bunt (Tilletia sp.) were believed to be mist, wind and excess water in the soil. The epidemics of late blight (Phytophthora infestans) that occurred on potato in Ireland in 1844 and 1845 were attributed to cool, rainy weather. More recently, Jones (1924), Gaumann (1950) and Yarwood (1959) pointed out the importance of environmental and other external factors on the possibility of plants to become diseased.

With the twentieth century a major branch of study in plant pathology began to focus on the influence of environmental factors on disease development. Studies on the interaction of components of the disease triangle have gradually evolved. Early observations centered around interactions between plant yield from bulk populations of crops species, disease epidemics incited by composite variants of a given parasite, and environment mostly referred to diverse altitude, latitude or season. In recent times, such studies have become more sophisticated and include interactions between host plants which may differ in one, two or more genes, strains or specific inoculum concentrations of the pathogen and specific environmental components, either climatic or nutritional. With the increasing contemporary need for introducing resistance to diversified types of microrganisms in an increased number of crop plants, simpler new screening methods are needed to check large populations of plant accessions and breeding progenies for resistance. In addition to this, the manner of inheritance of new resistance factor(s) to viral, fungal, bacterial and non-parasitic disease can be better investigated and fully understood when tests are carried out under controlled conditions.

The influence of endogenous and exogenous factors on the screening for oligogenic and polygenic resistance in tomato is here illustrated by the examples of Fusarium wilt and tomato-TMV.

3.1. Fusarium wilt

In addition to the effects of genetic plant constitution and the virulence of the parasite, wilting symptoms caused by F. oxysporum f. sp. lycopersici in tomato are influenced by such factors as availability of mineral elements, soil pH, temperature, humidity and light. Fusarium wilt of tomato is considered a warm-weather disease. In northern regions the importance of this disease is limited by low temperatures, except in greenhouses. Edgerton and Moreland (1920) were the first to show that Fusarium wilt develops more severely under higher greenhouse temperatures. Clayton (1923 and 1923b) observed that the optimum temperature for disease expression is about 28°C. Forester and Walker (1947) demonstrated that tomato plants are more susceptible to disease if they are kept at temperatures above or below the optimum for plant growth before inoculation. These Authors also observed that susceptibility to disease increased in plants grown in low humidity, low levels of nitrogen and phosphorus, high potassium levels, low soil pH, short photoperiodism and low light intensity. A detailed review of the effect of some environmental factors and mineral nutrients on disease development was carried out by Walker (1971).

In breeding for resistance, knowledge of the interaction between genotype and environment is necessary in order to differentiate between susceptible and resistant plants. Polygenic resistance in plants is often greatly affected by environment and plant age and in many cases its phenotypic expression is not detected because of the severe screening conditions to which plants are subjected during testing, i.e. high inoculum density, poor light intensity, low or high temperatures and plants which are too young. Polygenic resistance, in tomatoes as well as in many other species, can only be recovered under natural conditions of growth in adult plants and is, thus, also referred to as 'field' or 'mature plant' resistance. Screening techniques for selecting plants polygenically resistant to F. oxysporum f. sp. lycopersici in tomato at the early stage of growth were recently studied by Cirulli and Ciccarese (1982). In their greenhouse tests, the authors used plants at the two- and four- leaf stage, three inoculum dilutions of 10^1, 10^3 and 10^6 conidia/ml, and two temperature regimes of 22°C and 26°C. In general, under those conditions, the cultivar Super Marmande (possessing no resistance) was highly susceptible, the cultivar Bari RVF-SMD (possessing the I-1 gene) was highly resistant, and Marglobe and Rutgers (polygenically resistant) were slightly less susceptible than Super Marmande. This reaction pattern was not clearly delineated in Super Marmande, Marglobe and Rutgers when an inoculum dilution of 10^1 was used.

On the whole the mean reaction of Super Marmande, Marglobe and Rutgers varied in accordance with the inoculum dilution. Percentages of diseased plants and disease cultivar index significantly increased with increasing inoculum density, while disease incidence and severity indices were also significantly higher at 26°C than at 22°C. The percentage of diseased plants varied significantly according of plant age and was higher in plants inoculated at the four-leaf stage (Table VI). These results seem to exclude the use of inoculum at a dilution of 10^1 conidia/ml for 'early' recovery of polygenically resistant plants of "Marglobe" and "Rutgers" since 40-50% for plants at the two-leaf stage and 50-10% of plants at the four-leaf stage of the susceptible Super Marmande had escaped infection even 33 days after inoculation. The higher overall level of disease incidence observed in plants inoculated at the four-leaf stage is difficult to interpret, since plant defence mechanisms commonly increase with plant age. It may be possible that the root system of the four-leaf stage plant picks up more inoculum during inoculation than the less-developed roots of plants at the two-leaf stage (Table VII).

Table VII - Influence of inoculum diluition, temperature and plant stage on Fusarium wilt incidence (From: Cirulli and Ciccarese, 1980).

Inoculum dilution	Diseased plants (%)	Severity index (%)
10^6	81	56
10^3	73	48
10^1	36	20
Temperature regimes (C°)		
26	67	56
22	60	27
Plant stage		
4-leaf	71	44
2-leaf	56	39

Figures are means of disease records taken at 12, 19, 26 and 33 days after inoculation on the susceptible Super Marmande and on the polygenically resistant Marglobe and Rutgers.

"Marglobe" and "Rutgers" were commonly used until 1940 as source of Fusarium resistance, and the identification of resistant segregants was made on adult plants in the field. Since the discovery of I-1 and I-2 genes, the detection of Fusarium wilt resistance has been carried out in glasshouses using screening methods involving severe inoculation procedures. Although this allows for rapid identification of I-1 gene resistant plants, it does not permit the selection of polygenically resistant plants that are inevitably eliminated. The appearance of new races of F. oxysporum f.s. lycopersici could necessitate the use of polygenically resistant tomato germplasms or the combination of polygenic and monogenic resistance. Although the latter solution could reduce the risk of new disease outbreaks, it involves more complex breeding techniques, although these could be partially simplified by producing F_1 hybrids. The work of Cirulli and Ciccarese (1982) indicates that polygenically resistant plants can also be screened in greenhouses at the seedling stage using, as in the case of Rutgers-type of resistance, two-leaf stage seedlings, 10^3 conidium/ml, and either 22° or 26°C temperature regimes.

3.2. Tomato mosaic virus - TMV

Cirulli and Alexander (1969) found that the phenotypic expression of Tm-2^a was influenced by strains of TMV and temperature. These autors made their tests at two temperature regimes, namely 16-17°C and 26-28°C and used five strains of TMV. Temperature and virus strain conditioned the expression of disease symptoms. At 16-17°C two classes of plants, healthy and mottled, were observed after inoculation. At 26-28°C three classes of plants, healthy, systemically necrotic, and mottled were produced in inoculated segregating F_1, F_2, F_3 and BC generations indicating that resistance was controlled by a single dominant gene. At high temperatures, a reversal of dominance occurred, with heterozygous resistant plants exhibiting systemic necrosis (Figure 10).

Cirulli and Ciccarese (1975) carried out studies to clarify to interaction between TMV isolates, temperature, allelic conditions and combinations of the resistance genes Tm-1, Tm-2 and Tm-2^a, as well as to classify the TMV isolates on the basis of virus-strain/host-genotype relationships. They made their tests at 17°, 22°, 26° and 30°C. At all temperature regimes the percentage of virulent isolates was higher on heterozygous than on homozygous Tm hosts. The activity of each of the three Tm genes was temperature dependent, while with both heterozygous and homozygous allelic conditions, the percentage of isolates that exhibited virulence was higher at temperatures of 26° and 30°C than at 17° and 22°C. This gene reaction seems to suggest that this is a case of cumulative action, with two alleles for resistance producing twice the effect of one allele. Cirulli and Ciccarese (1975) found the temperature of 26°C and the use of Tm-gene homozygous differentials to be most satisfactory for strain classification of TMV (Table VI).

Vander Plank (1978) proposed a protein-for-protein hypothesis in

disease in which host and pathogen are involved gene-for-gene. He postulated that susceptibility involves the co-pôlymerization of protein from the
host and protein from the pathogen. In the gene-for-gene diseases, host
and pathogen recognized each other by their protein and protein polymeriza-
tion which is a reversible endothermic process carried out at rather high
temperatures. Vander Plank (1978) stated that the results by Cirulli and
Ciccarese (1975) are in complete agreement with this hypothesis.

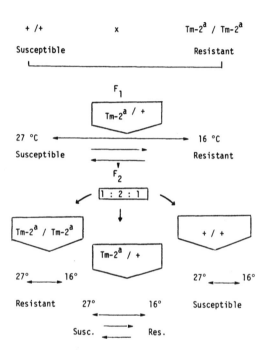

Figure 10 - Synotipic diagram of the influence of temperature and allelic condition on the resistance Tm-2^a
gene response to TMV.

In a further study using near-isogenic lines of tomatoes, Ciccarese
and Cirulli (1980) ascertained the influence of light intensity, inoculum
dilution, plant age and allelic condition on the expression of the Tm-1,
Tm-2 and Tm-2^a resistance genes. They tested three isolates of TMV,
different levels of light intensity (reduced greenhouse light ca. 7,000 lux;
normal greenhouse light ca. 14,000 lux; and greenhouse light + supplementary
artificial light ca. 30,000 lux at 11 a.m.), inoculum dilution (1:30, 1:900,
1:9,000, and 1:90,000 TMV-infected White Burley tobacco sap) and plant
growth 2-, 4-, and 6-leaf stages). In Tm-host/TMV-strain in homologous
combinations (i.e. those combination that in a preliminary test produced a
susceptible reaction in all plants grown in normal greenhouse light and

inoculated with a 1:30 TMV inoculum diluition) highest plant infections occurred at the 14,000 and 30,000 lux light intensity regimes. No symptoms
or very low percentages of diseased plants occurred with reduced light
intensity (7,000 lux). Variations of light intensity also induced changes in
symptoms: the systemic necrotic syndrome that at 7,000 lux with certain
Tm-2/+:TMV-isolate or Tm-2^a/+:TMV-isolate combinations occurred, was

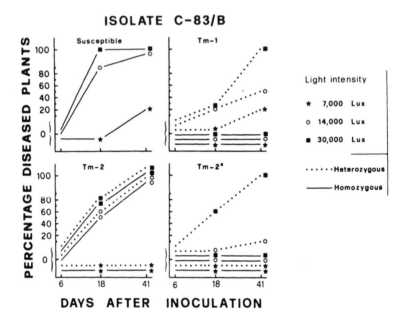

ISOLATE C-83/B

Figure 11 - Influence of light intensity and allelic condition
on the expression of susceptible and Tm-resistant near-iso
genic tomatoes.
In preliminary tests isolate C-88 was highly pathogenic on
the susceptible (possessing no Tm genes),Tm-1/+and Tm-2
host (mosaic symptom), and slightly active on Tm-1 / Tm-1
and Tm-2^a/+ tomatoes (symptom of mosaic and systemic necrosis on a few plants respectively).
From Ciccarese and Cirulli, 1980.

partially replaced at 14,000 lux and totally replaced at 30,000 lux by
systemic chlorotic spotting. Moreover, on susceptible plants the shortest
incubation periods occurred with highest levels of inoculum (Figure 11).
Plant age had no effect on disease frequency in the susceptible tomato line
whereas in homologous Tm-Host/TMV-strain combinations the percentage of
diseased plants diminished with increasing plant age (Figure 12).

These studies definitively showed that the tomato/TMV system is markedly affected by the exogenous factors of light intensity, temperature regimes and inoculum dilution as well as by the endogenous factors of

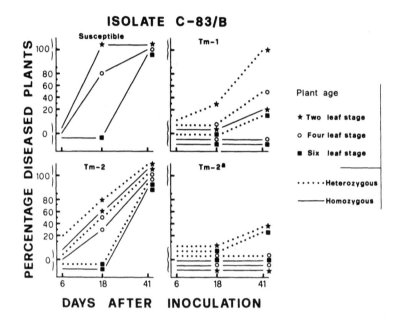

Figure 12 - Influence of plant age and allelic condition on infection of near-isogenic susceptible and Tm-resistant tomatoes. *From Ciccarese and Cirulli, 1980.*

plant age and allelic condition and combination of the Tm resistance genes. Consequently, these findings should be taken into account in TMV resistance studies and in the routine work of plant breeding when uniformity of disease development, shortening of screening time and lowering of the risk of escape are required.

4. Classification of tomato pathogens and tomato disease resistance

Disease resistance in tomato can be classified into two types (Cirulli, 1974): 1-oligogenic which is highly effective against specific races of pathogens and is inherited in a discontinuous mendelian manner: it includes I-1 and I-2 genes for resistance fo Fusarium wilt; Ve gene for resistance to Verticillium wilt, Cfn for Cladosporium fulvum; Ph-1 and Ph-2 for Phytophthora infestans; Sm for Stemphylium solani; Pto for Pseudomonas syringae pv tomato; Tm-1, Tm-2 and Tm-2ᵃ for mosaic-TMV; 2-polygenic, which is effective against a large spectrum of specific races of the pathogen and is inherited as continuous, quantitative segregation: it includes resistance to Alternaria solani, Colletotrichum phomoides, Phoma destructiva, Sclerotium rolfsii, Septoria lycopersici, Corynebacterium michiganense and Pseudomonas solanaceum (Table VIII).

Table VIII – Characterization of tomato oligogenic and polygenic disease resistance.

Characteristics	Resistance	
	Oligogenic	Polygenic
Reaction to indifferentiated originary population of the pathogens	High level	Uniform, intermediate levels
Reaction to several races of the pathogen	High against some races, but not operating against other races of the pathogen	Uniform, intermediate levels against all races
Stability in time	Variable with the type of pathogen	Commonly unchanging
Effects of environment on resistance	Commonly, none	Commonly affected by temperature and plant age
Screening for resistance at seedling stage	Effective	Ineffective
Screening for resistance at adult stage	Effective	Effective
Examples of pathogens against which resistance has been found	*Verticillium* *Fusarium* Mosaic-TMV *Meloidogyne* *Pseudomonas syringae pv tomato*	*Alternaria solani* *Colletotrichum coccodes* *Fusarium* *Corynebacterium michiganense*

In tomato, the capability of the pathogens to produce new races matching new resistant cultivars depends on: the kind of sporogenesis, the type and number of resting structures, the mode of dissemination, the mechanism of penetration into the host and the type of parasitism. In relation to these characteristics, tomato pathogens can be classified (Table IX) into two groups (Cirulli, 1974): Type A, comprising those with an infrequent

Table IX - Tomato pathogens classified according to the rate of appearance of new races.

Pathogens		
Characteristics	Type A	Type B
Dissemination of spores or propagules	Short distance	Long distance
Number of spores or propagules	Small	Large
Frequency of new races	Low	High
Examples of pathogens:	*Fusarium* *Verticillium* *Pyrenochaeta* *Meloidogyne*	*C. fulvum* *P. infestans* Mosaic-TMV

appearance of new races (i.e. Fusarium, Verticillium, Meloidogyne) and Type B which includes pathogens that frequently mutate to new races (i.e. Cladosporium fulvum, Phytophthora infestans, Tomato mosaic virus-TMV).

References

Alexander L.J. (1934). Leaf mold resistance in the tomato. Ohio Agr. Exp. Stn. Wooster, Ohio, 26 pp.

Alexander L.J. (1962). Susceptibility of certain Verticillium resistant tomato varieties to an Ohio isolate of the fungus. Phytopathology, 52, 998-1000.

Alexander L.J. (1963). Transfer of a dominant type of resistance to the four known Ohio pathogenic strains of tobacco mosaic virus (TMV), from Lycopersicon peruvianum to L.esculentum. Phytopathology, 53, 869 (Abstr.).

Andrus C.F. and G.B Reynard (1945). Resistance to Septoria leaf spot and its inheritance in tomatoes. Phytopathology, 35, 16-24.

Andrus C.F., G.B. Reynard, H. Jorgensen and J. Eades (1942). Collar rot resistance in tomatoes. Jour. Agr. Res., 65, 339-346.

Andrus C.F., G.B. Reynard and B.L. Wade (1942). Relative resistance of tomato varieties, selections and crosses to defoliation by Alternaria and Stemphylium. U.S.D.A. Circ. 652, 23 pp.

Bailey D.M. (1940). The seedling test method for root-knot nematode resistance. Proc. Amer. Soc. Hort. Sci. 38, 573-575.

Bailey D.M. (1947). The development through breeding of greenhouse tomato varieties resistant to leaf mould. Rep. hort. Stn. Ont. for 1945-46, pp. 56-60.

Barham W.S. and N.N. Winstead (1957). Inheritance of resistance to root-nematodes in tomatoes. Proc. Amer. Soc. hort. Sci., 69, 372-377.

Barksdale T.H. (1974). Evaluation of tomato fruit for resistance to Rhizoctonia soil rot. Pl. Dis. Reptr., 58, 406-408.

Barksdale T.H. and A.K. Stoner (1973). Segregation for horizontal resistance to tomato early blight. Pl. Dis. Reptr., 57, 964-965.

Barksdale T.H. and A.K. Stoner (1977). A study for inheritance of tomato early blight resistance. Pl. Dis. Reptr., 61, 63-65.

Barksdale T.H. and A.K. Stoner (1978). Resistance in tomato to Septoria lycopersici. Pl. Dis. Reptr., 62, 844-847.

Bashi E., M. Pilowsky and J. Rotem (1973). Resistance in tomato to Stemphylium floridanum and S. botryosum f. sp. lycopersici. Phytopathology, 63, 1542-1544.

Bender C.G. and P.B. Shoemaker (1977). Prevalence and severity to Verticillium wilt of tomato and virulence of Verticillium dahliae Kleb. isolates in Western North Carolina. Proc. Amer. Phytopathol. Soc., 4, 152 (Abstr.).

Bohn G.W. and C.M. Tucker (1940). Studies on Fusarium wilt of the tomato. I. Immunity in Lycopersicon pimpinellifolium Mill. and its inheritance in hybrids. Missouri Agr. Exp. Stn. Res. Bull., 311, 82 pp.

Bonde R. (1929). Physiological strains of Alternaria solani. Phytopathology, 19, 533-548.

Bonde R. and E.F. Murphy (1952). Resistance of certain tomato varieties and crosses to late blight. Maine Agr. Exp. Sta. Bull. No. 497, 15 pp.

Boukema I.W. (1982). Inheritance of resistance to Didymella lycopersici Kleb. in tomato (Lycopersicon Mill.). Euphytica, 31, 981-989.

Bouhot D., P. Erard and P. Camporota (1974). Apparition en France de la race 2 de la fusariose vasculaire de la tomate. Annales de Phytopathologie, 10, 485-487.

Broadbent L. (1964). Tomato mosaic. Rep. Glasshouse Crops. Res. Inst., 90-91.

Campbell R.N., V.H. Schweers and D.H. Hall (1982). Corky root of tomato in California caused by Pyrenochaeta lycopersici and control by soil fumigation. Pl. Dis. Reptr., 66, 657-661.

Ciccarese F. and M. Cirulli (1980). Influence of light intensity, inoculum dilution and plant age on the expression of Tm resistance genes in tomato. Phytopath. Z., 98, 237-245.

Ciccarese F. and M. Cirulli (1983). Comportamento di cultivar e linee di pomodoro verso la suberosità radicale. Inf.tore fitopatol., XXXIII, n.11, 57-59.

Cirulli M. (1968). Ricerca di fonti di resistenza verso tre malattie del pomodoro in specie di Lycopersicon. Ann. Fac. Agr. Univ. Bari, 22, 361-371.

Cirulli M. (1969). Un isolato di Verticillium dahliae Kleb. virulento verso varietà resistenti di Pomodoro. Phytopath. medit., 6, 132-136.

Cirulli M. and L.J. Alexander (1966). A comparison of pathogenic isolates of Fusarium oxysporum f. s. lycopersici and different sources of resistance in tomato. Phytopathology, 56, 1301-1304.

Cirulli M. and L.J. Alexander (1969). Influence of temperature and strains of tobacco mosaic virus on resistance in a tomato breeding line derived from Lycopersicon peruvianum. Phytopathology, 59, 1287-1297.

Cirulli M. and F. Ciccarese (1975). Interaction between TMV isolate, temperature, allelic condition and combination of the Tm resistance genes in tomato. Phytopath. medit., 14, 100-105.

Cirulli M. and F. Ciccarese (1980). Influenza di alcuni fattori sull'espressione delle resistenze poligenica e monogenica a Fusarium oxysporum f.s. lycopersici nel Pomodoro. Inf.tore fitopatol., 11, 41-45.

Cirulli M. and F. Ciccarese (1982). Factors affecting early screening of tomatoes for monogenic and polygenic resistance to Fusarium wilt. Crop Protection, 1, 341-348.

Clayton E.E. (1923). The relation of temperature to the Fusarium wilt of the tomato. Amer. J. Botany, X, 71-87.

Clayton E.E. (1923a). The relation of soil moisture to the Fusarium wilt of the tomato. Amer. J. Botany, X, 133-146.

Collingwood E.F. and M. Defranco (1979). Fusarium oxysporum f.s. lycopersici pathotype 2 on tomato. FAO Plant Protection Bulletin, 27, 22-23.

Doolittle S.P. (1929). Soil transmission of Tomato mosaic and streak in the greenhouse. Phytopathology, 18, 155.

Ebben Marion H., M.J.W. Smith and E.A. Turner (1978). Tolerance of tomatoes to Pyrenochaeta lycopersicy: comparison of a tolerant line with a susceptible cultivar in infested soils. Pl. Path., 27, 91-96.

Edgerton C.W.and C.C. Moreland (1920). Tomato wilt. La. Agr. Exp. Sta. Bull., 174.

Ellis D.E. (1943). Root knot resistance in Lycopersicon peruvianum. Pl. Dis. Reptr., 27, 402-404.

Foster R.E. and J.C. Walker (1947). Predisposition of tomato to Fusarium wilt. J. of Agr. Res., 74, 165-185.

Frazier W.A. and R.K. Dennet (1949). Tomato line of Lycopersicon esculentum type resistant to tobacco mosaic virus. Am. Soc. Hort. Sci. Proc., 54, 265-271.

Gabe H.L. (1975). Standardization of nomenclature for pathogenic races of Fusarium oxysporum f.s.lycopersici. Trans. Brit. Mycol. Soc., 64, 156-159.

Gaümann E. (1950). Principles of Plant infection. Transl. W.B. Brierley. London: Crosby Lockwood and Son Ltd. 543 p.

Gallegly M.E.(1952). Physiologic races of the tomato late blight fungus. Phytopathology, 42, 461-462.

Gallegly M.E.(1964). West Virginia '63, a new home-garden tomato resistant to late blight. Bull. West Va. Agric. Exp. Stn. Sci., 490.

Gallegly M.E. and M.E. Marvel (1955). Inheritance of resistance to tomato race 0 of Phytophthora infestans. Phytopathology, 45, 103-109.

Gilbert J.C. and D.C. McGuire (1956). Inheritance of resistance to severe root-knot from Meloydogine incognita in commercial type tomatoes.Proc. Amer. Soc. hort. Sci., 68, 437-442.

Goode M. (1966). New race of tomato Fusarium wilt fungus. Ark. Farm. Res., 15(1), 12.

Graham K.M. (1955). Distribution of physiological races of Phythopthora infestans (Mont.) de Bary in Canada. Am. Potato J., 32, 277-282.

Graham K.M. and T.C. Yap (1976). Studies on bacterial wilt. I. Inheritance of resistance to Pseudomonas solanacearum in Tomato. Malaysian Agr. Res., 5, 1-8.

Grattidge R. (1982). Screening for resistance to a third race of Fusarium wilt in tomato. Australasian Plant Pathology, 11, 29-30.

Grattidge R.and R.G. O'Brien (1982). Occurrence of a third race of Fusarium wilt of tomatoes in Queensland. Plant Dis., 66, 165-166.

Grogan R.G., K.A. Kimble and I. Misaghi (1975). A stem canker disease of tomato caused by Alternaria alternata f.sp. lycopersici. Phytopathology, 65, 880-886.

Hall D.H. and K.A. Kimble (1972). An isolate of Verticillium found pathogenic to wilt-resistant tomatoes. Calif. Agric., 26(9), 3.

Hendrix J.W. and W.A. Frazier (1949). Studies on the inheritance of Stemphylium resistance in tomatoes. Hawaii Agric. Exp. Stn. Tech. Bull., 8, 24 p.

Henning R.G. and L.J. Alexander (1952). Evidence of existence of physiologic races of Alternaria solani. Phytopathology, 42, 467.

Hogenboom N.G. (1970). Inheritance of resistance to corky root in tomato (Lycopersicon esculentum Mill.). Euphytica, 19, 413-425.

Holmes F.O.(1954). Inheritance of resistance to infection by tobacco-mosaic virus in tomato. Phytopathology, 44, 640-642.

Holz G.(1976). Race two of Fusarium oxysporum f. lycopersici in the Republic of South Africa. Phytothylactica, 8, 87-88.

Hubbeling N.(1968). Attack of hitherto resistant tomato varieties by a new race of Cladosporium fulvum and resistance against it. Meded. Rijksfacult. Landb. Wetensh. Gent, 31, 1011-1017.

Hubbeling N., L.J. Alexander and M. Cirulli (1971). Resistance to Fusarium and Verticillium wilts in tomato. Meded. Fakult. Landb. Wetensch. Gent, 36, 1006-1016.

Jones L.R. (1924). The relation of environment to disease in plants. Amer. J. Bot., 11, 601-609.

Jong J.De and S. Homma (1976). Inheritance of resistance to Corynebacterium michiganense in the tomato. Journal of Heredity, 67, 79-84.

Kanwar J.S., E.A. Kerr and P.M. Harney (1980). Linkage of Cf-1 to Cf-11 genes for resistance to tomato leaf mold, Cladosporium fulvum Cke. Report Tomato Genetics Cooperative, No. 30, 20-21.

Kanwar J.S., E.A. Kerr and P.M. Harney (1980a). Linkage of Cf-12 to Cf-24 genes for resistance to tomato leaf mold, Cladosporium fulvum Cke. Report Tomato Genetics Cooperative, No. 30, 22.

Katan J. and I. Wahl (1966). Occurrence in Israel of new dangerous isolates of the Fusarium wilt pathogen. Proc.First Congr.Medit. Phytopathol. Union, 425-430.

Kerr E.A. and D.L. Bailey (1964). Resistance to Cladosporium fulvum Cke. obtained from wild species of tomato. Canad. J. Bot., 42, 1541-1554.

Kurozawa C. and M.A. Pavan (1982). Distribution of physiologic races of Fusarium oxysporum f.sp. lycopersici (WR) Snyder et Hansen in San Paulo State. Summa Phytopathologica, 8, 153-160.

Laterrot H. (1972). Une résistance à la maladie des racines liégenses chez la tomate. Annales de l'Amèlioration des Plantes, 22, 109-113.

Laterrot H. (1983). La lutte génétique contre le maladie des racines liégenses de la tomate. Revue Horticole, No 238 , 23-35.

Laterrot H. (1983a). Reoccurrence of Pt o. Report Tomato Genetics Cooperative, No 33, 2-3.

Laterrot H. (1984). Specific resistance of Verticillium dahliae race 2 in tomato. Report Tomato Genetics Cooperative, No 34, 10-11.

Laterrot H. and P.Pecaut (1966). Presence of race 2 of Fusarium oxysporum f. sp. lycopersici (Sacc.) Snyd. and Hansen in tomato crops in Morocco. Proc. First Cong. Medit. Phytopathol. Union, Bari, Italy, 431-433.

Lesley J.W. and M. Shapovalov (1937). The Riverside tomato. A new variety resistant to two wilt diseases. Seed World, March 26, 1937.

Liu C.Y., S.T. Hsu and M.L. Ho (1974). Breeding tomato for resistance to Pseudomonas solanacearum. I. Intervarietal differentiation in resistance to Ps. solanacearum. From Plant Breeding Abstract, 46, 11609.

McFarlane J.C., E. Hartzler and W.A. Frazier (1946). Breeding tomatoes for nematode resistance and for high vitamin C content in Hawaii. Proc. Amer. Soc. hort. Sci., 47, 262-271.

Miller R.E. and D.L. Kananen (1968). Occurrence of Fusarium oxysporum f. sp. lycopersici race 2 causing wilt of tomato in New Jersey. Pl. Dis. Reptr., 52, 553-554.

Moore W.D. and G.B. Reynard (1945). Varietal resistance of tomato seedlings to the stem lesion-phase of Alternaria solani. Phytopathology, 35, 933-935.

O'Brien R.G. and D. G. Hatton (1981). Identification of race 2 of Verticillium wilt in tomato in south-east Queensland. Australasian Plant Pathology, 10, 56-58.

Okie W.R. and R.G. Gardner (1982). Breeding for resistance to Verticillium dahliae race 2 of tomato in North Carolina. J. Amer. Soc. hort. Sci., 107(4), 552-555.

Patrick Z.A., E.A. Kerr and O.L. Bailey (1971). Two races of Cladosporium fulvum new to Ontario and further studies of Cf1 resistance in tomato cultivars. Can. J. Bot., 49, 189-193.

Person C. (1959). Gene-for-gene relationships in host: parasite system. Can. J. Bot., 37, 1101-1130.

Pilowsky M. and S. Cohen (1974). Inheritance of resistance to tomato yellow leaf curl virus in tomatoes. Phytopathology, 64, 632-635.

Pilowsky M. and D. Zutra (1982). Screening wild tomatoes for resistance to Bacterial speck pathogen (Pseudomonas tomato). Plant Dis., 66, 46-47.

Pilowsky M. and D. Zutra (1982a). Evidence for the presence of different genes for resistance to Pseudomonas tomato in Lycopersicon esculentum cv. Ontario 7710 and in L. pimpinellifolium accession P.I. 126430. Report Tomato Genetics Cooperative, No. 32, 42.

Pitblado R.E. and B.H. Mac Neill (1983). The genetic basis of resistance to Pseudomonas syringe pv. tomato in field tomatoes. Can. J. Plant Pathol., 5, 251-255.

Pritchard F.J. (1922). Development of wilt-resistant tomatoes. U.S.D.A. Bull. 1015, Washington.

Pitchard F.J. (1975). Marglobe tomato. Market Growers Journal, 40, 104-105.

Reynard G.B. and C.F. Andrus (1945). Inheritance of resistance to the collar-rot phase of Alternaria solani in tomato. Phytopathology, 35, 25-26.

Riggs R.D. and N.N. Winstead (1959). Studies on resistance in Tomato to root-knot nematodes and on the occurrence of pathogenic biotypes. Phytopathology, 48, 716-724.

Romshe F.A. (1942). Nematode resistance test of tomatoes. Proc. Amer. Soc . hort. Sci., 40, 423.

Sasser J.N. (1954). Identification and host-parasite relationships of certain root-knot nematodes (Meloidogyne spp.). Md. Agr. Expt. Sta. Tech. Bull. A-77, 31 pp.

Schaible L., O.S. Cannon and V. Waddoups (1951). Inheritance of resistance to Verticillium wilt in tomato cross. Phytopathology, 41, 986-990.

Sengbush R.V. and N. Loschakowa-Hasenbush (1932). Immunitätszüchtung bei Tomaten. Vorläufiges über die Züchtung gegen die Braunfleckenkrankheit (Cladosporium fluvum Cooke) resistenter Sorten. Der Zuchter, 4, 257-264.

Shapovalov M. and J.W. Lesley (1940). Wilt resistance of Reverside variety of tomato to both Fusarium and Verticillium wilts. Phytopathology, 30, 760-768.

Smith P.G. (1944). Embryo culture of a tomato species hybrid. Proc. Amer. Soc. hort. Sci., 44, 413-416.

Sonoda R.M., J.J. Augustine and R.B. Volin (1980). Bacterial wilt of tomato in Florida: history, status and sources of resistance. Proc. Fla. State hort. Society, 92, 100-102.

Soost R.K. (1963). Hybrid tomato resistant to tobacco mosaic virus. Inheritance for resistance in derivatives of a complex species hybrid. J. Hered., 54, 241-244.

Stall R.E. (1961). Development of Fusarium wilt on resistant varieties of tomato caused by a strain different from race 1 isolates of Fusarium oxysporum f. sp. lycopersici. Pl. Dis. Reptr., 45, 12-15.

Stall R.E. and J.M. Walter (1965). Selection and inheritance of resistance in tomato to isolates of race 1 and 2 of the Fusarium wilt organism. Phytopathology, 55, 1213-1215.

Taylor A.L. and B.G. Chitwood (1951). Root-knot susceptibility of Lycopersicon peruvianum. Pl. Dis. Reptr., 25, 97.

Taylor A.L., C.E. Cox, A.L. Harrison, D.C. McGuire and F.A. Romshe (1955). Root-knot nematode disease. Meloydogine species. In: Disease resistance in wild species of tomato, North Central Res. Publ. 51 and Ohio Agr. Exp. Sta. Res. Bull., 732.

Thomason I.J. and P.G. Smith (1957). Resistance in tomato to Meloidogyne javanica and M. incognita acrita. Pl. Dis. Reptr., 41, 180-181.

Thyr B.D. (1972). Virulence of Corynebacterium michiganense isolates on Lycopersicon accessions. Phytopathology, 62, 1082-1084.

Thyr B.D. (1976). Inheritance of resistance to Corynebacterium michiganense in tomato. Phytopathology, 66, 1116-1119.

Tillet M. (1775). Dissertation sur la cause qui corrompt et noircit les grains de blé dans les épis; et sur les moyens de prévenir ces accidents. Bordeaux: Brun. (in Phytopathol. Classics No 5, 1937; 91 pp).

Tjamos E.C. (1981). Occurrence of race 2 of Verticillium dahliae in Greece. Annales Institut Phytopathologique Benaki, 12, 216-226.

Tokeshi H., F. Galli and C. Kurozawa (1966). Nova raça de Fusarium do tomateiro en Sao Paulo. Anais. Esc. sup. Agric. 'Luiz Queiroz' 23, 271-227. (Abstr., Rev. Appl. Mycol., 47, 166).

Turkensteen L.J. (1973). Partial resistance of tomato against Phytophthora infestans, the late blight fungus. Mededelingen, Institut voor Plantenziektenkunding Onderzoeg No 633, 88 pp.

Vakalounakis D.J. (1983). Evaluation of tomato cultivars for resistance to Alternaria blight. Ann. appl. Biol., 102, (Suppl.) 138-139.

Van der Plank J.E. (1978). Genetic and molecular basis of plant pathogenesis. Springer-Verlag; Berlin, Heidelberg, New York, 167 pp.

Volin R.B. and J.P. Jones (1982). A new race of Fusarium wilt of tomato in Florida and sources and resistance. Proc. Fla. State Hort. Society, 95, 268-270.

Volin R.B. and R.T. Jr Mc Millan (1978). Inheritance of resistance to Pyrenochaeta lycopersici in tomato. Euphytica, 27, 75-79.

Walker, J.C. (1971). Fusarium wilt of tomato. Americ. Phytopathol. Soc. Monograph No.6, 56 pp.

Watts V.M. (1947). The use of Lycopersicon peruvianum as a source of nematode resistance in tomatoes. Proc. Amer. Soc. hort. Sci., 49, 233-234.

Werner R.A., D.C. Sanders and W.R. Henderson (1980). Inheritance of tolerance to Rhizoctonia fruit rot of tomato. J. Amer. Soc. Hort. Sci., 105(6), 819-822.

Winstead N.N. and W.S. Barham (1957). Inheritance of resistance in tomato to root-knot nematode. Phytopathology, 47, 37-38.

Winstead N.N. and R.D. Riggs (1963). Stability of pathogenicity of B biotype of the root-knot nematode Meloidogyne incognita on tomato. Pl. Dis. Reptr., 47, 870-871.

Yarwood C.E. (1959). Predisposition. (In: Plant Pathology, ed. J.G. Horsfall and A.E. Dimond, vol. 1, 521-562. New York, London: Academic Press, 674 pp.).

Session 4
A: Other crops
B: General aspects on pest control in field vegetables

Chairman: C.Pelerents

Population dynamics of aphids as vectors of potato viruses in Portugal

M.O.Cruz de Boelpaepe & M.I.Rodrigues
Centro Nacional de Protecção da Produção Agrícola, INIA-ER, Tapada da Ajuda, Lisboa, Portugal

Summary

Studies were conducted in different sites of two major potato production areas of Trás-os-Montes: Montalegre and Boticas, ca. 44 km NW and 24 km SW of Chaves, respectively. Influence of climatic conditions between areas as well as of altitude and wood protection within each area on aphid attack was evaluated. Comparison of capture means for the key vector, Myzus persicae (Sulz.), was made between sites of each region. Flight dynamics of green peach aphid, M. persicae, was studied in untreated fields from several localities of Montalegre and Boticas. Seasonal abundance of aphidophagous insects was evaluated on potatoes. To predict flight activity, correlations between logarithm of the number of alatae M. persicae and accumulated day-degrees above 5°C were established. Relationship between number of alate aphids and virus incidence was analysed. On the basis of our results, the most appropriate timing and aphid density at which to apply insecticidal sprays were determined.

1. INTRODUCTION

The main aphid vectors of potato viruses are the green peach aphid, Myzus persicae (Sulz.), the buckthorn aphid, Aphis nasturtii (Kltb.), the foxglove aphid, Aulacorthum solani (Kltb.), the potato aphid, Macrosiphum euphorbiae (Thomas) (6) and Aphis frangulae (Kltb.) (4, 12).

M. persicae is, economically, the most important pest because of its efficiency on potato leaf roll virus (PLRV) spread. The probability of virus transmission is closely related to green peach aphid density and intensity of migration (5). Regarding the non-persistent potato virus Y (PVY) the efficacy of its transmission successively decreases from M. persicae (70%), A. nasturtii (20%), A. frangulae (14%), A. solani (14%) to M. euphorbiae (8%) (10).

Regular surveys of those species are necessary in potato fields to avoid substantial yield losses. Until 1979, abundance and flight activity of aphid vectors had not been evaluated in seed potato production areas of northern Portugal. For this reason, prospections of aphid species were initiated in several sites of Montalegre and Boticas. These mountain regions differ in climatic conditions.

Our assays conducted, from 1979 to 1981, in both regions had the objective to know seasonal fluctuations in the flight activity of aphid vectors. The influence of ecogeographical conditions was also studied. On the basis of green peach aphid's behaviour and virus incidence, a preliminary economic threshold for pest control was estimated.

257

2. MATERIAL AND METHODS

Studies were conducted in two regions of seed potato production in northern Portugal: Montalegre and Boticas, ca. 44 km NW and 24 km SW of Chaves, respectively. Montalegre is located in Larouco Mountain (altitude 1 525 m) and Boticas in the proximity of Barroso Mountain (altitude 1 261 m).

Annual mean temperatures and annual rainfall are 10°C and 1 400 mm in Montalegre and 12.5°C and 1 000 mm in Boticas. The speed of predominant winds, from N and W, is comprised between 6 and 20 km/h (14). Several sites differing by altitude or wood protection were prospected in each region to evaluate aphid abundance.

In Montalegre, at a height over 800 m (900-1 050 m) two fields located at Cambezes and Serraquinhos, respectively, had a wood protection against predominant winds. This wood protection was composed of Quercus robur L. associated with Pinus alepensis L. At the same altitude, two other fields, situated at Vilar de Perdizes and Firvidas, had no wood protection. Below 800 m (550-600 m) two fields chosen at Paradela and Loivos, respectively, were protected at N and W by association of Quercus robur L., Castanea sativa L. and Pinus pinaster L.. In contrast, there was no wood protection at Parada and Outeiro sites.

In Boticas, above 800 m (900-1 050 m), two fields located at Viveiro and Carvalho were protected against predominant winds by a Quercus robur and Castanea sativa association. At the same altitude, Alturas do Barroso and Lavradas sites were not protected. Below 800 m (650-700 m), Bosto Frio and Vilar presented at N and W dispersed trees of Quercus robur associated with Castanea sativa, in contrast with Quintas and Carvalhelhos sites without wood protection.

Flight activity of aphid vectors was followed in each field by means of two yellow water traps (31 x 24 x 9.5 cm) set up on the ground, near the corners of the field at 1.5 m beyond the planting. The alate aphids trapped were identified as to species and recorded weekly. Apterous aphids and immature stages of aphidophagous insects were counted on an upper, a middle, and a lower leaf from 50 randomised plants taken on the two diagonals of the field. Aphid counts on potatoes were made at the time of virus inspections.

Assessment of virus incidence (PLRV, PVY and mosaics) was based on the presence of leaf symptoms detected by visual process. Inspections on 100 randomised plants selected from the field diagonals were made when plants reached 15-20 cm height and at middle flowering (50% open flowers). Yield virus infection was determined a posteriori by preculture of 100 tubers collected per variety.

The selected fields were untreated against aphids, but received only one insecticidal application to control Colorado potato beetle (Leptinotarsa decemlineata Say.).

The numbers of alate green peach aphids, captured in several sites of each region, were compared by analysis of variance and significant means separated by Newman-Keuls multiple range test. Correlations between logarithm of the numbers of M. persicae alatae and accumulated day-degrees above 5°C (D°), calculated from a biofix point (beginning of the second flight) were established for Montalegre and Boticas regions.

3. RESULTS

3.1 Ecogeographical Influence of Altitude and Wood Protection on the Alate Green Peach Aphid Levels

For each region, comparison of the mean numbers of alate green peach aphids was made between several sites differing by altitude and/or wood protection against predominant winds (Table 1). Analysis of variance revealed no significant differences among all the sites of Montalegre in 1979 (F = 0.33, P > 0.05 for df_1 = 3 and df_2 = 28), 1980 (F = 2.07, P > 0.05 for df_1 = 3 and df_2 = 46) and 1981 (F = 1.15, P > 0.05 for df_1 = 3 and df_2 = 52). The same was found in the Boticas region in 1979 (F = 1.06, P > 0.05 for df_1 = 3 and df_2 = 30) and 1981 (F = 1.04, P > 0.05 for df_1 = 3 and df_2 = 28). Thus, the aphid levels are similar between 600 and 1 050 m of altitude and between fields, at the same height, with or without wood protection.

3.2 Flight Activity of the Green Peach Aphid (1979-1981) in the Montalegre Region

Flight curves were constructed from the logarithms of the mean numbers of alate aphids trapped in all the sites prospected in this region.

M. persicae had a bimodal flight periodicity in 1979, 1980 and 1981, with a peak in numbers in late spring (mid-June) or early July and another in late summer (Figure 1).

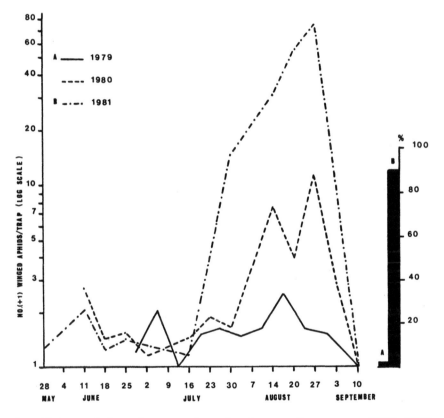

Fig. 1. Flight Activity of M. persicae in the Montalegre Area (1979-1981) and Total Virus Incidence on Plants (A) and Tubers (B) of Kennebec Potato Variety in 1979 and 1981, respectively

TABLE 1 Mean Numbers ($\pm$ SE) per Trap[b] of Alata Green Peach Aphids in Several Locations of Montalegre and Boticas

Fields[a]	Montalegre			Boticas	
	1979	1980	1981	1979	1981
800 m ^ A	0.53 ± 0.14 a	1.10 ± 0.79 a	21.96 ± 9.15 a	1.13 ± 0.47 a	9.00 ± 5.28 a
800 m ^ B	1.03 ± 0.58 a	4.00 ± 1.85 a	19.29 ± 10.66 a	0.29 ± 0.09 a	2.38 ± 1.30 a
800 m v A	0.69 ± 0.25 a	1.28 ± 0.64 a	5.01 ± 2.19 a	1.94 ± 0.77 a	2.31 ± 1.27 a
800 m v B	0.79 ± 0.19 a	0.58 ± 0.19 a	8.21 ± 4.43 a	1.28 ± 0.60 a	8.03 ± 3.76 a

a Ecogeographical conditions - A: with a wood protection against predominant winds (N and W); B: without wood protection.

b Means, within a column, followed by the same letter are not significantly different at the $P = 0.05$ level by analysis of variance.

In this mountain region, the winter is very severe. The daily mean temperatures until February are below the threshold of green peach aphid development (5°C). Under these unfavourable conditions, the species is most probably dioecious, with host alternation between a woody primary host, Prunus persica L. (Batsch), and an herbaceous secondary host.

Spring migration of the alatae from the woody host to potato crop could occur at the beginning of June, when daily maximum temperatures rose above 15°C (Table 2). That is the threshold of flight for the majority of aphid species (9, 12). This migration took place in 1979, 1980 and 1981 after 269.05, 281.90 and 263.90 D°, accumulated from 1 February, the date from which hatching of the eggs most probably occurred. Although the daily maximum temperatures in June 1979 were propitious to the flight, first alatae were not captured because water traps had been placed too late.

First flight had on each year a small amplitude with a peak reaching 1 alate per trap in 1979 and 1981, and 1.7 alatae per trap in 1980. The highest peak of the first flight can be explained by the higher average temperatures observed during April 1980, in comparison with those of 1979 and 1981. After the aphid literature (3, 12), the intensity of this flight depends on the April temperatures. First migrants, in regions where holocyclic clones prevail, are not dangerous because green peach aphids come from the primary host and are virus free (3).

Decline of the first flight, from the peak to mid-July, was more rapid in 1979 and 1980. The heavy rains which occurred in 1979 (10-12 July) and in 1980 (5-15 June) were the main cause of the drop in aphid levels.

A new increase in alate aphids (second flight) began between early and mid-July and reached the maximum peak in mid-August 1979 or in late August 1980 and 1981. During this dispersal flight, aphid levels increased more rapidly than those of the first flight. The highest alate level (75.2 alatae/trap) was observed in 1981 and the lowest (1.4 alatae/trap) in 1979 (Figure 1).

Adverse conditions of the potato host plant that reached maturity and unfavourable maximum temperatures in late August (near 30°C) were the most important factors responsible for aphid decline.

The outbreak of the aphid dispersal flight in 1981 can be mainly explained by the increase in summer temperatures (Table 2). During this warm period, aphids were likely to be carried from the valley to mountain by air convection currents. The second flight is very dangerous because of spreading potato viruses from crop to crop. In 1979, when aphid levels were below 2 alatae per trap, the percentage of virus infection (1.75% plants infected), assessed on plants of Kennebec potato variety, was lower than the allowable damage level for class B of seed potatoes (3% for the two combined inspections). In contrast, the higher levels of alatae found in 1981 caused serious damage in the yield of the same cultivar (90% tubers infected) (2).

3.3   Influence of Climatic Conditions on the Activity of Green Peach Aphid
       Dispersal Flight

The influence of temperature and rainfall on the dispersal flight activity was studied over the three years (1979-1981).

During population growth, the logarithms of alate numbers (+ 1) linearly increased with accumulated day-degrees above 5°C, calculated from the beginning of the dispersal flight (Figure 2). Significant correlations were obtained in 1979 ($r = 0.860$, $t = 3.37$, $P < 0.05$), 1980 ($r = 0.922$, $t = 5.32$, $P < 0.01$) and 1981 ($r = 0.970$, $t = 8.85$, $P < 0.001$).

261

TABLE 2 Average of Daily Maximum and Mean Temperatures (± SE) and Total Rainfall (mm) in Montalegre and Boticas

Year/Month	Montalegre Max.Temp.(°C)	Montalegre Mean Temp.	Montalegre Rainfall	Boticas Max.temp.(°C)	Boticas Mean temp.	Boticas Rainfall
1979						
April	10.36 ± 0.74	6.33 ± 0.48	215.0 (18d)	—	—	—
May 1-15	19.15 ± 1.42	12.61 ± 1.33	109.8 (14d)	—	—	—
16-31	10.89 ± 0.83	7.43 ± 0.73		—	—	—
June	21.87 ± 0.71	15.61 ± 0.77	2.5 (3d)	—	—	—
July	24.60 ± 0.66	18.54 ± 0.53	49.2 (4d)	26.09 ± 0.76	18.08 ± 0.51	0.2 (1d)
August	23.82 ± 0.65	18.05 ± 0.56	6.0 (5d)	22.67 ± 0.49	16.40 ± 0.39	0.0
1980						
April	13.30 ± 0.66	7.97 ± 0.45	77.9 (6d)	—	—	—
May 1-15	11.62 ± 0.61	7.88 ± 0.57	78.8 (12d)	—	—	—
16-31	15.33 ± 0.58	10.31 ± 0.15		—	—	—
June	17.56 ± 0.79	13.10 ± 0.57	51.5 (8d)	—	—	—
July	22.25 ± 0.82	16.07 ± 0.60	7.7 (9d)	—	—	—
August	24.78 ± 0.72	18.84 ± 0.67	20.5 (5d)	—	—	—
1981						
April	11.22 ± 0.53	7.02 ± 0.45	85.5 (19d)	16.89 ± 0.50	11.39 ± 0.42	104.4 (26d)
May 1-15	11.81 ± 1.14	7.73 ± 0.79	134.6 (22d)	18.04 ± 0.99	12.68 ± 0.65	73.2 (24d)
16-31	13.28 ± 0.89	9.59 ± 0.71		19.70 ± 0.78	13.79 ± 0.73	
June	22.12 ± 1.18	15.86 ± 0.82	25.3 (6d)	28.09 ± 1.06	20.26 ± 0.73	11.6 (3d)
July	24.44 ± 0.83	18.14 ± 0.71	1.8 (3d)	30.65 ± 0.82	21.74 ± 0.65	7.8 (2d)
August	26.23 ± 0.60	19.91 ± 0.50	3.7 (3d)	31.84 ± 0.87	22.81 ± 0.38	0.0

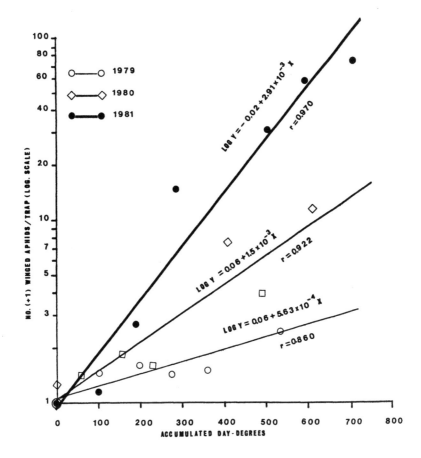

Fig. 2. Linear Relationships between Logarithm of the Green Peach Aphid
Number (+ 1) per Trap and Accumulated Day-degrees Above 5°C
Calculated from the Beginning of the Second Flight, in the
Montalegre Area (1979-1981)

Due to the fact that aphid levels varied considerably from year to
year, tests for parallelism demonstrated the slopes of regression lines for
the three years were significantly different ($F = 47.13$, $P < 0.05$ for
$df_1 = 2$ and $df_2 = 14$). Thus a common regression line by pooling over years
cannot be constructed. The correlations above evidenced the determinant role
played by summer temperatures, ranging from 15 to 30°C (12), in the
increasing dispersal flight.

Rainfall influence on the dispersal flight was generally inconspicuous
during the summer period (Table 2). Weak rains occurring between mid-July
and late August (Figure 3) did not affect alate levels in 1979 and 1981. In
contrast, rainfalls in mid and late August 1980 contributed to a temporary
reduction in aphid numbers (Figure 1).

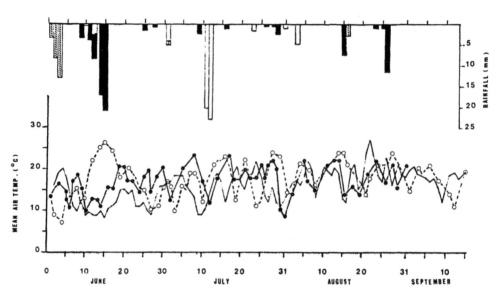

Fig. 3. Daily Mean Air Temperatures and Rainfall in Montalegre (1979-1981)

3.4 Flight Pattern of the Other Aphid Vectors in Montalegre

Flight curves of all of the aphids A. frangulae, A. nasturtii, A. solani and M. euphorbiae showed an initial higher peak tending to decline through great fluctuations in aphid numbers (Figure 4). This pattern of flight reflected the unimodal periodicity which appears to be characteristic of these species (13).

The fluctuating decline is a consequence of a non-coincidence in the dates of the species migrations. Migrants of A. frangulae, A. nasturtii and A. solani reached their maximum in the potato crop between late June and early July and the M. euphorbiae ones between early July and mid-August.

The peak of the total number of these aphid vectors varied between 6.4 and 39.1 aphids per trap in 1979 and 1980, respectively. The predominance of the initial peak is related to the anholocyclic behaviour of the vectors that overwinter on herbaceous hosts. Thus, first migrants arrived earlier and the intensity of this spring flight is generally greater than that of the holocyclic species (3). Differences in the intensity of the initial peak in 1979, 1980 and 1981 were due to April temperatures (Table 2). The alatae were more abundant in 1980 because temperatures in April were higher.

Although this species group is less effective than M. persicae in virus transmission, its abundance and infectiousness during spring migration are responsible for the crop contamination.

The vectors A. frangulae, A. nasturtii, A. solani and M. euphorbiae represented 37.6 to 43.4% of the total number of aphids captured in 1979 and 1981 respectively. Relative abundance of the key vector M. persicae in the traps varied from 8.9 to 26.0% during the same period.

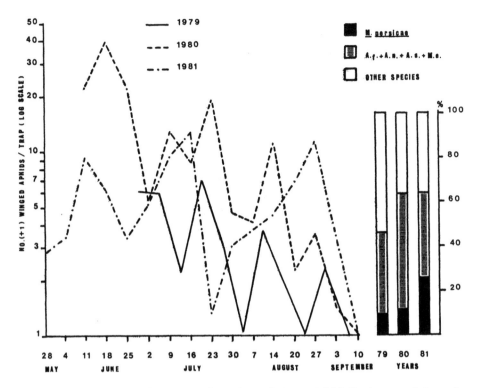

Fig. 4. Fluctuations of the Total Number of Some Aphid Vectors and Annual
Relative Abundance of this Species Group in Yellow Water Traps in
the Montalegre Region.
(Abbreviations: A.f. - Aphid frangulae; A.n. - Aphis nasturtii;
A.s. - Aulacorthum solani; M.e. - Macrosiphum euphorbiae

Annual mean number of each aphid vector per trap was not significantly
different from 1979 to 1981 for A. nasturtii, A. solani and M. euphorbiae.
In contrast, A. frangulae was more abundant in 1980 than in 1979 while
M. persicae reached the highest density in 1981 (Table 3).

3.5 Predators and Parasites of the Aphid Populations Infesting Potato
Fields in Montalegre (1979)

Two counts of aphids and associated parasites and predators were made
in 1979 on potatoes of several fields located in the Montalegre region. As
shown in Table 5, parasitism plays the most important role in the reduction
of aphid density at the initial phase of population growth (first count).
The maximum rate of parasitism was observed in mid-July and reached 65.0%.
Parasite species were not identified.
With respect to predator insects, only the eggs of chrysopids
(Neuroptera) were counted on the crop. Reproductive potential of these
predators was higher in late July. From this time, chrysopids concurred to
reduce pest incidence on plant potatoes. Chrysoperla carnea Stephens was the
predominant lacewing identified in field samples.

265

TABLE 3 Annual Mean Numbers (± SE) per trap^a of Alate Aphids Captured in Montalegre and Boticas

Species	Montalegre			Boticas	
	1979	1980	1981	1979	1981
A. frangulae	1.25 ± 0.14a	9.48 ± 2.97b	3.01 ± 0.79ab	0.55 ± 0.08a	2.04 ± 0.59a
A. nasturtii	1.05 ± 0.26a	2.68 ± 0.75a	1.44 ± 0.24a	0.38 ± 0.16a	0.39 ± 0.10a
A. solani	0.06 ± 0.03	0.12 ± 0.08	0.02 ± 0.01	0.02 ± 0.01	0.09 ± 0.06
M. euphorbiae	1.98 ± 0.40a	1.10 ± 0.43a	1.57 ± 0.26a	5.41 ± 2.38a	1.09 ± 0.13a
M. persicae	0.79 ± 0.09a	1.74 ± 0.66a	13.62 ± 3.58b	1.17 ± 0.29a	5.43 ± 1.55a

^{a} For each region, means within a row with letters in common are not significantly different at the P = 0.05 level by Newman-Keuls' multiple range test.

The behaviour of green peach aphid in Boticas was identical to that observed in Montalegre, with a bimodal flight periodicity (Figure 5).

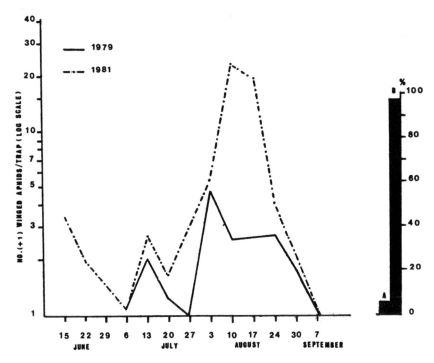

Fig. 5. Flight Activity of M. persicae in the Boticas Region and Total Virus Incidence on Plants (A) and Tubers (B) of Désirée Variety in 1979 and 1981 respectively

Migration of first alatae to the potato crop could occur from the beginning of April, when the daily maximum temperatures rose above the threshold of flight. At this time, there was 216.5 D°, accumulated from 1 February. Thus, first flight in Boticas should be more precocious than in the Montalegre region. However, first flight took place later in 1981 because of the heavy and continuous rains until mid-June (Table 2). Accumulated day-degrees were not calculated for 1979 because there was no available meteorological data.

Dispersal flight began between early and late July, increasing faster than in the Montalegre region. Its peak reached 3.6 and 22.3 alatae per trap in 1979 and 1981, respectively.

Decline of alate number was observed from the beginning of August in both years. The drastic fall in the aphid numbers seems due to the high maximum temperatures in August, which daily mean (31.84 ± 0.87) became unfavourable to survival of the alatae (12).

Virus incidence on plants of Désirée potato variety in 1979 reached the level of 6.2% in the combined inspections. A virus outbreak in 1981 (Figure 5) caused 97.0% tubers to be infected in the same cultivar (2).

3.7 Influence of Climatic Conditions on the Activity of Green Peach Aphid
     Dispersal Flight

     As regards temperature influence on the dispersal flight, a
significant correlation (r = 0.927, t = 4.955, P < 0.01) was found in 1981,
during the alate increase (from 6 July to 3 August), between logarithm of
the alate number (+ 1) and accumulated day-degrees above 5°C (Figure 6). The
same model of relationship was constructed for the Montalegre region. These
facts confirm that amplitude of the dispersal flight depends on the
accumulated temperatures during population growth.

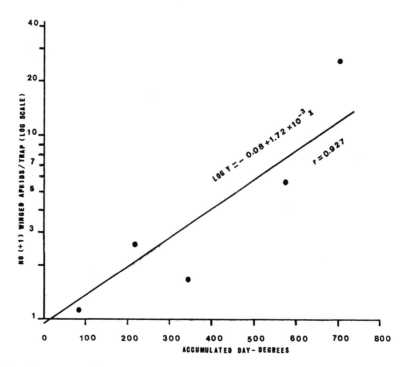

Fig. 6. Linear Relationships Between Logarithm of the Green Peach Aphid
        Number (+ 1) per Trap and Accumulated Day-degrees Above 5°C
        Calculated from the Beginning of the Second Flight, in the Boticas
        Area (1981)

3.8 Flight Pattern of the Other Vectors in Boticas

     Flight behaviour of the aphid vectors A. frangulae, A. nasturtii,
A. solani and M. euphorbiae in the Boticas region was identical to that
observed in the Montalegre area, particularly in 1981. In that year, the
upper peak (9.2 alatae per trap) observed in mid-June was followed by a
fluctuating decline, reflecting the non-coincidence of the unimodal flight
activity of those species. The magnitude of this flight was lower (Figure 7)
than in the Montalegre area.
     The flight curve in 1979 gave a better idea of the unimodal flight
periodicity, but the maximum peak (9.3 alatae per trap) was displaced to

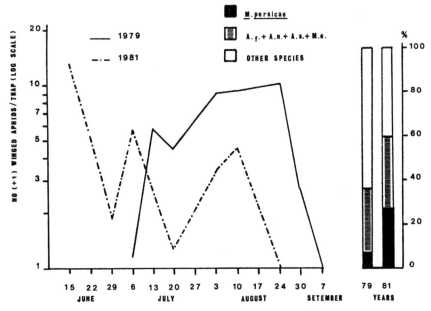

Fig. 7. Fluctuations of the Total Number of Some Aphid Vectors
(A. frangulae, A. nasturtii, A. solani and M. euphorbiae) and Annual
Relative Abundance of this Species Group in Yellow Water Traps in
the Boticas Region (1979 and 1981)
(Abbreviations: A.f. - Aphis frangulae; A.n. - Aphis nasturtii;
A.s - Aulacorthum solani; M.e. - Macrosiphum euphorbiae)

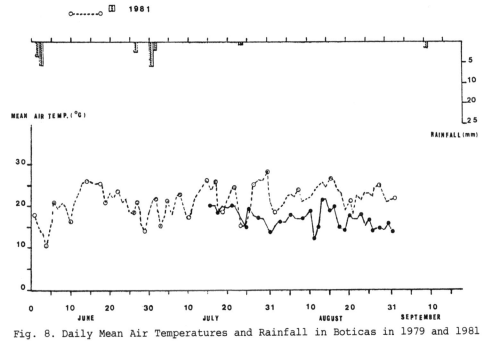

Fig. 8. Daily Mean Air Temperatures and Rainfall in Boticas in 1979 and 1981

269

TABLE 5 Aphid Density and Attack Rate by Parasites and Predators in 1979

	Fields	Date	Total 'live' aphids per 100 leaves	% of aphids attacked by parasites	Predator/prey[a] ratio x 100	
					Cecidomyid larvae in sample	Chrysopid eggs in sample
Montalegre >800 m	A	12 July	2	50.0	–	0.0
		24 July	34	14.0	–	1.9
	B	5 July	36	15.0	–	0.0
		16 July	14	7.0	–	0.0
<800 m	A	12 July	28	65.0	–	9.5
		24 July	153	2.8	–	0.4
	B	5 July	35	0.0	–	0.0
		17 July	5	0.0	–	25.0
Bottcas >800 m	A	12 July	2	50.0	0.0	300.0
		14 August	8	62.5	8.3	0.0
	B	12 July	0	100	0.0	0.0
		14 August	12	45.4	0.0	27.7
<800 m	A	12 July	2	0.0	0.0	0.0
		14 August	124	0.0	0.0	0.0
	B	12 July	0	100	0.0	0.0
		14 August	142	10.1	3.3	1.3

[a] live prey

A - with a wood protection against predominant winds (N and W)

B - without wood protection

late August. This delay was likely related to July and August daily mean temperatures which were lower than those recorded in 1981 (Figure 8).

The vectors A. frangulae, A. nasturtii, A. solani and M. euphorbiae represented 29.4 and 33.2% of the total number of aphids captured in 1979 and 1981, respectively. The relative abundance of M. persicae in the traps reached 6.7 and 27.5% in 1979 and 1981 respectively (Figure 7).

Annual mean numbers of each aphid vector per trap were not significantly different in the years 1979 and 1981 (Table 3).

A list of the aphid species trapped in both regions (Boticas and Montalegre) is given in Table 4.

TABLE 4 List of Aphid Species Captured in Water Traps

---

1.   Main vectors of potato viruses

     Aphis frangulae Kltb.
     Aphis nasturtii Kltb.
     Aulacorthum solani (Kltb.)
     Macrosiphum euphorbiae (Thomas)
     Myzus persicae (Sulz.)

2.   Other species*

     Aphis citricola van der Goot
     Aphis craccivora Koch
     Aphis fabae* Scop.
     Brachycaudus helichrysi (Kltb.)
     Brevicoryne brassicae (L.)
     Cavariella aegopodii (Scop.)
     Cavariella theobaldi (Gill. et Bragg.)
     Hyperomyzus lactucae (L.)
     Myzocallis castanicola Baker
     Myzus ascalonicus* Donc.
     Myzus ornatus* Laing
     Rhopalosiphoninus latysiphon* (Davids.)
     Rhopalosiphum padi (L.)
     Sitobion avenae (Fab.)
     Sitobion fragariae (Walk.)

---

* species indicated by an asterisk can also transmit potato viruses.

3.9   Predators and Parasites of the Aphid Populations Infesting Potato Fields in Boticas (1979)

In the Boticas region, parasitism played an important role in the reduction of aphid numbers at the initial stage of the pest infestation. At that time (12 July), parasites killed between 50 and 100% of the individuals (Table 5). Later, when the aphid population reached its peak (mid-August), the effectiveness of parasites decreased.

Concerning predation, only larvae of cecidomyids and eggs of lacewings (Chrysoperla carnea Stephens) were recorded in aphid samples. The predator/prey ratio for cecidomyids was higher in mid-August, when aphid numbers rose to the peak. In mid-July, chrysopids had the highest reproductive potential. From this period, their larvae contributed with parasites to reduce aphid density (Table 5).

271

4. DISCUSSION AND CONCLUSIONS

Dynamics of the aphid vectors of potato viruses particularly emphasises the behaviour of the alatae because of their effectiveness on virus spread.

Our studies on the flight activity of M. persicae, conducted in several localities of Montalegre and Boticas, demonstrated for both regions that the mean aphid numbers were not significantly different between 600 and 1 050 m of altitude. This result agrees with the author's findings concerning the behaviour of the same aphid vectors at different altitudes in Saint Michael Island, Azores (1). After Hurst (8), when maximum temperatures are higher and higher (summer), convection currents carry M. persicae to a height of 1 500 m. This explains why the variation of altitude mentioned above was not an obstacle to the arrival of the pest during the warmest months.

The inefficacy of the woods against aphid invasions was likely due to the dispersion of the trees.

The dispersal flight of M. persicae occurred more precociously in Boticas. This precocity seems to be related to the temperatures of the last part of June (13). In Boticas, the daily maximum temperatures in June were effectively greater than in Montalegre. Dispersal flight reached maximum magnitude in Montalegre where summer temperatures were more favourable to the flight.

Analysis of the flight activity of M. persicae and total virus incidence showed damage levels were reached as soon as aphid numbers rose above 2 alatae per trap. In Boticas, a mean number of 3.5 alatae per trap in 1979 was associated to a damage level (6.2% plants infected by viruses). The highest numbers of alatae in 1981 were responsible for outbreaks of virus disease in both regions.

On the basis of Montalegre and Boticas data, the preliminary economic threshold is two green peach aphids per trap and per week. Above this level, insecticidal applications were necessary.

Concerning A. frangulae, A. nasturtii, A. solani and M. euphorbiae, their maximum peak of activity occurred at the initial stage of the crop growth. These species coming from herbaceous hosts are generally infected by viruses. For this reason, a preventive insecticidal application is required to avoid crop contamination.

Although parasites and predators did not provide acceptable control of virus vectors, their importance should not be underestimated. Parasites were particularly active from the beginning of aphid infestations, which agrees with Hughues (7) and Robert's (11) findings.

Use of selective insecticides harmless to beneficial arthropods is recommended.

ACKNOWLEDGEMENTS

We would like to thank our colleagues Mr. Delfim Santos, Mrs. Dulce Anastácio and their co-workers of the Regional Direction of Trás-os-Montes e Alto Douro for their help in the completion of this study. We also thank Mrs. Isabel Gonçalves, Mrs. Olimpia Bicho and Mrs. Rosa de Jesus (National Centre for Crop Protection) for their technical assistance. We are grateful to Dr. Robert De Boelpaepe for the fruitful advice given in the revision of the manuscript.

REFERENCES

(1) CRUZ DE BOELPAEPE, M.O., 1982. Dinâmica sazonal dos afídeos
     virulíferos da batateira. Correlações com os factores abióticos. II
     Cong. Int. Soc. port. Ent., Funchal, Madeira, 28 September -
     3 October. 1981, Bolm. Soc. port. Ent. 25: 20-21 (Abstract)

(2) DIVISÃO DE BATATA-SEMENTE, Ex-DGPPA, INIA-ER, 1984. Projecto de apoio
     à batate-semente.

(3) DIXON, A.F.G., 1973. Biology of aphids. The Institute of Biology.
     Studies in biology No.44. Edward Arnold (Publishers) Limited: 1-58.

(4) GABRIEL, W. et al., 1972. Die Beziehungen zwischen dem Auftreten der
     Vektoren und der Höhe des Virusbesatzes (Blattroll-und Y-Virus) bei
     Kartoffeln in Mitteleuropa. Ziemniak: 27-60.

(5) GALECKA, B. and KAJAK, A., 1971. Studies on ecological mechanisms
     reducing population of Myzus persicae (Sulz.) (Hom. Aphididae). Ekol.
     Pol. 19: 789-806.

(6) HODGSON, W.A., POND, D.D., and MUNRO, J., 1942. Diseases and pests of
     potatoes. Information Services, Agriculture Canada, Ottawa: 1-69.

(7) HUGHUES, R.D., 1963. Population dynamics of the cabbage aphid
     Brevicoryne brassicae (L.). J. Anim. Ecol. 32: 393-423.

(8) HURST, G.W., 1969. Les aspects météorologiques des migrations des
     insectes. Endeavour 28: 77-81.

(9) JADOT, R., 1976. Aspects des épidémies de jaunisse et de la mosaïque
     de la betterave. I. Symptômes, modes de transmission, méthodes
     d'identification, vecteurs. Revue de l'Agriculture 3: 555-582.

(10) PROESELER, G. and WEIDLING, H., 1975. Die Retentionszeit von Stämmen
     des Kartoffel-Y-virus in verschiedenen Aphidenarten und Einfluss des
     Temperatur. Arch. Phytopathol. U. Pflanzenschutz Berlin 11: 335-345.

(11) ROBERT, Y., 1979. Recherches écologiques sur les pucerons Aulacorthum
     solani Kltb., Macrosiphum euphorbiae Thomas et Myzus persicae Sulz.
     dans l'Ouest de la France. III. Importance du parasitisme par
     Hyménoptères Aphidiidae et par Entomophtora sur pomme de terre. Ann.
     Zool. Ecol. anim. 11 (3): 371-388.

(12) ROBERT, Y., 1981. Pucerons de la pomme de terre. Journées d'études et
     d'information: Les pucerons des cultures. Paris, 2-4 March 1981. ACTA:
     195-212.

(13) ROBERT, Y. and ROUZE-JOUAN, J., 1978. Recherches écologiques sur les
     pucerons Aulacorthum solani Kltb., Macrosiphum euphorbiae Thomas et
     Myzus persicae Sulz. dans l'Ouest de la France. I. Etude de l'activité
     de vol de 1967 à 1976 en culture de pomme de terre. Ann. Zool. Ecol.
     anim. 10: 171-185.

(14) SERVIÇO NACIONAL DE METEOROLOGIA 1974. Atlas climático de Portugal.

Susceptibility of various bean cultivars to bacterial blights

C.G.Panagopoulos & D.A.Biris

Laboratory of Plant Pathology, Athens College of Agricultural Sciences and Plant Protection Institute, Volos, Greece

Bacterial blights (Common blight, Fuscous blight, and Halo blight) caused by Xanthomonas campestris pv. phaseoli (syn. Xanthomonas phaseoli, X. phaseoli var. fuscans) and Pseudomonas syringae pv. phaseolicola (syn. Pseudomonas phaseolicola) are widely distributed in most regions growing beans in Greece. In some areas these seed-borne bacterial diseases have become endemic and when weather conditions are favorable the attacks may cause very considerable crop losses (up to 70% yield reduction), especially on certain cultivars.

In laboratory experiments 16 varieties and local populations of beans obtained from the Fodder Crops and Pastures Institute of Larissa, Greece, were tested for their reaction to these bacterial pathogens.

The test plants were grown in pots under artificial light (16 hr photoperiod) at a temperature of 18-30°C and were inoculated 15-30 days after sowing, when the first trifoliate leaves had appeared. The inoculum (bacterial suspension containing approximately 10^5 cells/ml was prepared with SDW from a 48h nutrient agar growth) was applied with a hand or mechanical sprayer to the lower surfaces of the primary and trifoliate leaves from a distance of about 10 cm and with a pressure sufficient to impregnate the tissues with the inoculum. Fifteen plants from each cultivar and each pathogen were inoculated and an equal number of control plants were treated with sterile water.

First symptoms appeared usually one week after inoculation as water-soaked spots, which rapidly turned to necrotic often with a yellow halo and could involve a large portion of the leaf blade. Other symptoms, that might appear later, were uneven development of the leaflets, cupping of the leaf blades, wilting of leaves, or even

stunting, withering amd death of the whole plant as a result of systemic infection. On the contrary, some of the impregnated tissues of
the check plants soon turned to necrotic, due to mechanical damage and
there was not any further development of symptoms.

The results were evaluated 20-30 days after inoculation, the whole
series of inoculations being repeated 2-3 times.

The varieties ΦΞ-4(M-10657), ΦΞ-5(M-10658), ΦΣ-35(Rapsani), ΦΣ-39,
ΦΣ-46(Pyrgetos), ΦΣ-47(Carla), ΦΣ-48(Lyda), ΦΣ-49(Myrsini), M-4403
(Aridea) and M-10633 (Mayorquais) and the local populations M-6537
and M-6538 (originated from Xanthi), M-6546 (from Orestias), M-6834
(from Imathia), M-7356(from Konitsa) and M-14215 (from Velestino)
were all found to be susceptible to all pathogens, without any perce-
ptible differences in susceptibility.

In some instances some check plants also exhibited blight symptoms,
which indicated that the seed was contaminated to some degree.

Management practices by the bean grower can substantially reduce
bacterial blights in beans. These include the following: Use of disease-free seed. Use a 2-3 year crop rotation. Plow under plant de-
bris as soon as possible after harvest. Avoid irrigation with sprin-
klers. Spray with antibiotics or coppers.

276

Integrated control of the artichoke moth *(Gortyna xanthenes* Ger.) (Lepidoptera: Noctuidae) in Italy

E.Tremblay & G.Rotundo
Istituto di Entomologia Agraria, Università di Napoli, Portici, Italy

Summary

The Noctuid moth Gortyna xanthenes Ger.,which is a key pest of artichoke in the West Mediterranean areas,has been studied for several years. The species shows one annual generation and a definite adult flight period between September and November. In most areas eggs undergo a short winter diapause and hatch in January or February of the next year. In other regions (e.g.Sardinia),where no egg diapause occurs,the larvae of the new generation emerge between the end of October and the beginning of December of the same year. During its activity the larva bores all parts of the plant up to flower heads (globes) and spoils them. Attempts of chemical control by farmers lead to the contamination of globes even at harvest time. The discovery of the sex pheromone blend of the species allowed the field evaluation of different types of sex traps.One of them (a metal can filled with a layer of motor oil on water above which the dispenser is suspended) proved to be the most effective. Destruction of infested plants and shifting of harvest can reduce pest attacks to minimal values without chemical action.

In Italy artichoke cultivation is carried out on 53,000 hectars distributed in the Central and Southern part of the country,including Sardinia and Sicily. The artichoke moth,Gortyna xanthenes Ger.,is a key pest in the West Mediterranean area (part of South Italy,Sicily,Sardinia,South France,Spain and North-West Africa)(2). Moths emerge from the woody roots of the plants between September and November (9). Females attract males and after copulation lay eggs on the soil in proximity of plants or even outside the field. In Italy,eggs do not hatch until the end of January- mid February,with the exception of Sardinia where no egg diapause occurs and hatching takes place between the end of October and the beginning of December (3). The newly-hatched larvae reach the host plant leaves and penetrate into the midrib usually from the apex. After having bored the leaf rib down to its basis,the larvae mine the flower stems, reach the flower heads (globes) and spoil them. The last-stage larvae become positively geotropic,return into the flower axes and by the end of spring perforate the woody roots of the plants where the pupate at the end of summer. Infestation of old plants may reach 100%,but direct damage to globes is generally lower and is correlated to the coincidence of flower heads with larvae ascending the flower stems. Natural enemies are present(4,11) but their action is of limited value. One of the causes of their lack of effectiveness may be the number of chemical sprays (4-5 on the average) which are carried out against this and a few other key pests of artichoke(5). The erratic behavior of newly-hatched larvae and graduality in hatching require field-wide sprays with persistent contacticides. These should be combined with systemic compounds to eliminate those larvae which have started to mine the leaf ribs.

During five years of investigation on this species we could analyse several possibilities of reducing chemical pressure at harvest. These possibilities are here presented and discussed.

1. Adult trapping

In 1982 and 1983 mass trapping of adults was attempted by means of light traps and of traps baited with virgin females(9). Results confirmed the occurrence of the single autumnal peak of moth flights in all areas under observation. The use of semisinthetic diets (6) allowed the rearing of the moth in the laboratory and the identification of four components of the sex pheromone of the species (Z-11-hexadecenal, Z-9-hexadecenal, hexadecanal and Z-11-hexadecenol) (7,8). Z-11-hexadecenol was found to act as calling inhibitor toward Heliothis armiger (Hb.) males. Roof-type traps were first used to test the attractants, but they proved to be inadequate to the large size of the moths.Double cone and Heliothis-type traps were also used. The best catching results were finally obtained by using metal can water traps filled with a layer of motor oil on water above which the dispenser is suspended(10). The effect of mass trapping and of the confusion technique are still in course of evaluation.

2. Shifting of harvest times

Artichoke globes reach maturity in different periods of

the year. Late maturing cultivars are able to escape severe winters because their globes are ready for market from
March to June.These varieties, however, receive the highest
direct damage because their flower heads emerge and mature
in coincidence with the ascending activity of moth larvae.
Early maturing varieties which are cultivated in warmer areas
can be harvested from November to March. They produce most
of their globes during the egg diapause period or when young
larval stages live as leafminers and thus almost escape moth
attack. In the Sardinian conditions, in which egg hatching
takes place in autumn, the adoption of late maturing cvs
would allow to carry out in October-November safe chemical
treatments against Gortyna and Depressaria young larvae.On
the other hand late cvs are not requested by the market in
Sardinia.

3. Sanitation

Artichoke fields are often allowed to produce globes for
five consecutive years. At the end of this long production
period woody roots become infested above 100% (i.e. several
larvae and many empty galleries in each root). In all cases
plants are cut off at soil level in June-July. Sanitation,
that is careful removal of all infested plants in full summer,in no cases interferes with harvest and can in theory
eliminate the infestation from the field.This is what really happens in Sardinia on several hundreds of hectars (S.Or-
tu.pers.comm.) where problems of soil humidity or high inco-
mes induce farmers to remove all plants in June and to re-im
plant each year at the end of summer.

4. Conclusion

In conclusion,the life cycle of this pest presents two
vulnerable points,i.e. (a) a well-delimited flight period,
when male trapping or the technique of confusion can be applied with success,and (b) localization of mature larvae and
pupae in plant roots in summer,at the end of harvest,when sanitation procedures become feasible. Shifting of harversting
times is counteracted by climatic (e.g.in the Salerno area)
or market (e.g.in Sardinia) problems. As a matter of fact
only in the warm area of Bari (Apulia) early cultivation is
possible,in advance of Gortyna attacks to flower heads.In
the other situations a combination of pheromone use and of
sanitation may eliminate the necessity of chemical control.
Chemical action should be permitted with short residual insecticides under supervised control only in definite cases
of key pest coexistence(1,3).

REFERENCES

1. DELRIO, G. (1985). Osservazioni sul comportamento degli adulti di Depressaria erinaceella Stgr.(Lep. Oecophoridae) su carciofo. Inf.Fitopat., 35, 31-34.

2. IPPOLITO, R. and PARENZAN, P. (1978). Contributo alla conoscenza delle Gortyna Ochs. europee (Lepidoptera, Noctuidae). Entomologica, Bari,14, 159-202.

3. PARENZAN, P., IPPOLITO,R., ORTU, S., ROTUNDO, G., SINACORI, A. and TREMBLAY, E. (1985). Dati comparati sulla bio-etologia della Gortyna xanthenes Germ. (Nottua del carciofo) in diverse aree dell'Italia meridionale e insulare. Inf.Fitopat., 35, 35-40.

4. PRIORE,R. and TREMBLAY,E. (1981). Pseudovipio castrator F. (Hymenoptera Braconidae) entomofagi di Gortyna xanthenes Germ.(Lepidoptera Noctuidae). Boll.Lab.Ent.Agr. "Filippo Silvestri", Portici,38, 275-281.

5. PROTA, R. (1985). La difesa del Carciofo dai parassiti animali. Inf.Fitopat., 35, 9-18.

6. ROTUNDO, G. and GIACOMETTI, R. (1981). L'allevamento di Gortyna xanthenes Germ. (Lepidoptera Noctuidae) su dieta semiartificiale. Boll.Lab.Ent.Agr."Filippo Silvestri",Portici, 38, 245-250.

7. ROTUNDO, G.,MASSARDO,P., TONINI,C. and PICCARDI, P.(1981). Il feromone sessuale di Gortyna xanthenes Germ. (Lep. Noctuidae) con trappole di diverso tipo. Boll. Lab.Ent.Agr."Filippo Silvestri",Portici, 38, 231-243.

8. ROTUNDO, G., PICCARDI, P. and TREMBLAY, E. (1982). Preliminary report on the artichoke moth (Gortyna xanthenes Ger.) sex pheromone and its possible use in integrated control. Les Médiateurs chimiques, Versailles 16-20 Nov. 1981. INRA Pub. (Les Colloques de l'INRA, 7),277-279.

9. ROTUNDO, G., IPPOLITO, R. and PARENZAN, P. (1983). Catture di Gortyna xanthenes Germar(Lepidoptera Noctuidae) con trappole di diverso tipo. Nota preliminare. Atti XIII Congr.Naz.Ital.Ent.,Sestriere-Torino, 245-252.

10. ROTUNDO, G. and TREMBLAY, E. (1985). Field evaluation of female sex pheromone of the artichoke moth, Gortyna xanthenes. Entom.exp.appl., 37, in press.

11. SINACORI,A. (1985). Osservazioni bioecologiche sulla Gortyna xanthenes (Germar) (Lep.:Noctuidae) in Sicilia. Atti XIV Congr.Naz.Ital.Ent.,Palermo,Erice,Bagheria, 523-524.

The researches have been supported by a grant from the Italian Ministry of Education (MPI 40%).

Insect pest management by intracrop diversity: Potential and limitations

T.H.Coaker

Department of Applied Biology, University of Cambridge, UK

Summary

Intercrop diversity or polycultures can result in higher crop yields
for a number of reasons of which one clearly demonstrated advantage
is that insect pest populations are greatly reduced. Ecological
hypotheses suggest two ways that can explain pest reduction in
polycultures, 1) from a greater number of predators and parasites
and 2) by disruption of the pests' movement and reproductive
behaviour, the visual and chemical stimuli from host and non-host
plants affect both the rate at which insects colonise habitats and
their behaviour in those habitats. Examples are presented from
experiments with annual crops to demonstrate ways of studying the
ecological mechanisms causing pest suppression by intercropping.

1.1 Background

Agriculturalists have been using traditional systems of multicropping,
that is, growing more than one crop on the same piece of land in one year,
for centuries, and intercropping or polyculture is one type of cultural
system that is a common feature in the developing world, particularly in
the tropics and sub-tropics. In general, intercropping is practiced
because it can result in higher crop yields for a number of reasons that
include reduced weed competition due to denser ground cover provided by
the crops, soil conservation, better use of incident radiation, water and
soil nutrients (Willey, 1979).

An additional demonstrated advantage is that insect populations are
often reduced compared with those found on similar crops grown in
monoculture (Perrin, 1977). Such multicropping systems are thought to
possess a great potential for improved crop productivity and provide "some
lessons that temperate agronomists could usefully learn" (Norton & Conway,
1977). Entomologists, therefore, have an important contribution to make
to the advancement of our knowledge on the merits of intercropping, but
over the past 30-40 years, since the advent of synthetic insecticides,
there has been a general disinterest in cultural methods of pest control
and particularly multicropping as they are counteractive to high input
monoculture crop production.

Although pesticides are invariably the only means of effective pest
control, it can be argued that routine cultural practices including
spacial and temporal arrangements of crops should be regarded as the base-
line for pest management programmes and in so doing, the tendency to
depend on one form of control, such as chemicals or plant resistance could
be avoided.

Another feature that distracted applied entomologists away from this
area of research was the belief that meaningful manipulation to agro-

ecosystem diversity had to be done on a large scale and on inter- rather
than intra- crop vegetation because of the well developed migratory and
host plant finding abilities of most phytophagous insects (Price &
Waldbauer, 1975). These concepts have been shown to be incorrect
(Dempster & Coaker, 1974) and clearly there is a greater value in
methods that can be applied to a crop in one field irrespective of what is
done nearby.

Another obstacle to its progress was the questioning of the widely
held assumption prior to the early 1970's that increased diversity lead to
increased stability as it was proposed at least in theory, that a complex
network of ecological interactions in more diverse systems could lead to
fluctuating insect populations, not more stable ones, and that there was
no general correlation between increased diversity and increased stability
in the sense of reducing insect outbreaks (van Emden & Williams, 1974).
Even the very logic of the original idea was questioned and these views
have lead to widespread scepticism about the value of using ecological
theory to suggest strategies in agricultural diversification for insect
pest control.

Much of the confusion in the debate derives from the range of
definitions of stability and diversity (Levins & Wilson, 1980), but in the
agricultural context reasonable concepts that might be acceptable are:
1) species or structural diversity can be achieved by adding different
plant species to monocultures, that is by intercropping with unrelated
plant taxa, or by allowing weeds to grow within the crop, and 2) that
where stability is defined to represent low fluctuations in insect
populations over time, this is not what agriculturalists are interested
in. Farmers require the pest to be maintained at a relatively low level
of abundance below an economic threshold acceptable to the production of
the crop and this allows for some fluctuations in pest numbers but below
that threshold (Risch et al., 1983).

The question then becomes whether by diversifying agroecosystems pest
populations are reduced below economic threshold and in consequence
results in economic benefits.

Two hypotheses attempt to explain why intercropping can reduce
phytophagous pest levels (Root, 1973). The first, the Enemies or
Predation hypothesis, predicts a greater number of insect natural enemies
in polycultures than monocultures which give a better suppression of pest
populations. Polycultures provide better conditions for natural enemies
by providing: 1) greater temporal and spacial distribution of nectar and
pollen sources which increase the numerical response of natural enemies,
2) increased ground cover, particularly important to noctural predators
and 3) increased prey that offer alternative food sources when the pest
species are scarce or at an appropriate time in the predators life cycle.
The second, the Resource Concentration hypothesis, implies that associated
plant species may have a direct effect on the ability of the herbivore to
find and utilize its host plants. It also involves the movement and
reproductive behaviour of the pest insect. Therefore, for any particular
pest species it is the total strength of the attractive stimuli which
determines the 'resource concentration' resulting from the different
interacting factors such as the density and spacial arrangement of the
host plant and the interfering effects from the non-host plants. So the
lower the 'resource concentration' the more difficult it will be for the
pest insect to locate a host plant. Relative 'resource concentration'
also increases the probability that the pest species will leave its
habitat once it has arrived, for instance, the pest may tend to fly sooner
and further after landing on a non-host than on a host plant which results

282

in a higher emigration rate from polycultures than monocultures. A
reduction in colonization and an increase in emigration obviously results
in fewer pests on their host plants. The question then to be asked is can
these hypotheses explain reported cases of pest reduction in diversified
systems and if so to below economic thresholds that result in economic
benefits?

A recent survey of 150 published studies on the effect of
diversifying agroecosystems on insect herbivores in which only intracrop
diversity, i.e. intercropping and weedy culture systems were included,
were classified into those in which herbivore populations were decreased
below that found in the simple system, those in which there was no change
and those when the populations were increased. Of the 198 species
observed in nearly 800 examples examined, 53% of the herbivores were
decreased, 18% increased, 20% showed a varied response and 9% showed no
change (Risch et al., 1983). Since these examples do not represent random
selections of herbivores and agroecosystems such aggregate percentages
must be treated with caution, nevertheless, these data highlight several
important and unanswered questions - one of which might be, did a decrease
in pest abundance lead to better crop yields? Unfortunately only 19 of
these 150 studies reported crop yields but their interpretation is
difficult, for example how should comparisons between polyculture and
monoculture yields be made? Recent opinion suggests that more diverse
systems usually yield better when compared on a Land Equivalent basis
(Mead & Willey, 1980) since these changes in yield could result from
changes in pest abundance as well as from the other agronomic variables
mentioned.

Another question arising from the survey concerns the ecological
mechanisms involved that accounted for the differences in pest abundance.
Unfortunately only about 10% of the case studies reported conducted
relevant experiments to reveal the mechanisms responsible for the observed
differences in herbivore abundance. It is clearly useful to have an
understanding of these mechanisms if we want to predict how plant
diversification affects pest populations and to explain why exceptions
occur. Knowledge of the mechanisms involved would allow finer tuning of
the agronomics of the crop mixture with respect to timing of planting,
relative crop growth rates and spacial arrangements. Amongst the studies
in the survey that attempted to evaluate the mechanisms involved only a
few authors compared the relative contribution of natural enemies and
herbivore movement patterns to account for herbivore reduction in
diversified systems. In these cases the impact of natural enemies was
less than pest emigration rates in polycultures suggesting that 'resource
concentration' was more important in suppressing pest numbers. A more
recent survey has shown, however, that from 228 published examples on 79
species of the effect of diversification of agroecosystems on predator
abundance, 48% (11 spp.) were higher in the diverse system, 13% (10 spp.)
shared no difference and 25% (70 spp.) showed a variable response. These
results clearly show that plant diversity often results in higher predator
populations (and presumably higher predation rates). On the other hand if
'resource concentration' is the main cause of herbivore abundance this can
be associated with current theoretical arguments concerning insect-plant
interactions based on the strategy of plants in disturbed habitats (e.g.
annual plants) to escape from herbivores in space and time (being hard to
find). This leads to insect pest outbreaks in annual monocultures because
the plants are more easily found compared with those grown in polycultures.

In conclusion, insect pest abundance is inclined to be suppressed by
intracrop diversity compared with simple cropping systems but whether this

causes increases in yield or not requires more critical experimentation. 'Resource concentration' may be equally if not more important than natural enemies in causing pest suppression.

1.2 Experimental tests
1.2.1 Brassicas
Observations on insect herbivore populations attacking brassicas in diverse plant communities were made initially on rape grown in weedy (diverse) and weed-free (simple) plots and then repeated by substituting the weeds with red clover (cv. Granta) to provide a more controllable non-host plant structural diversity. The main herbivore species found on the rape and their % reduction in the diverse compared with the simple systems are shown in Table 1. The % reduction varied between species and the early and late parts of the season due to the relative growth of the clover, the taller clover making the rape plants less apparent, i.e. more difficult to find by the immigrating insects.

Table 1. Percentage reduction of insect infestation on rape (Brassica oleracea) in diverse compared with simple plant communities

Species	May–July	August–September
Phyllotreta rufipes	59	62
Meligethes spp.	37	59
Ceutorhyncus assimilis	18	46
Mamestra brassicae	62	82
Phytomyza rufipes	58	62
Delia radicum	64	89
Aleyrodes proletella	-	92
Brevicoryne brassicae	83	76
Myzus persicae	25	58

These experiments were then followed by intercropping Brussels sprouts with clover (cv. Kersey White) which was either sown in autumn or early spring to provide 0, 25, 50 and 100% ground cover between the rows of sprouts when they were planted in late-April. One tenth hectare plots were used to reduce inter-treatment effects but this large plot size was found to be unnecessary as similar treatment effects were obtained on 10 metre square plots (O'Donnell & Coaker, 1975). Subsequent agronomic developments included early summer cauliflowers (cv. All Year Round) and summer cabbages (cv. Primo) intercropped with dwarf broad beans (cv. The Sutton)and French beans (cv. Blue Lake), respectively. The intercrops were sown before the brassicas so that their growth was equivalent to that of the brassicas at transplanting.

Pest suppression in each of these experiments was successful from the better polycultures; cabbage aphid, Brevicoryne brassicae, was reduced by up to 90%, cabbage root fly, Delia radicum, by up to 70% (Coaker, 1980; Tukahirwa & Coaker, 1982). Lepidoptera populations were low and were not assessed but Dempster and Coaker (1974) and Theunissen and Den Ouden (1980) reported reductions in their numbers from intercropping brassicas with clover and Spergula arvensis.

Crop yields compared by Land Equivalent Ratios (i.e. the relative land area required as a monoculture to produce the same yield as intercropping) showed that marketable yields of brassicas exceeded those grown in monoculture when untreated with insecticide, but were generally lower when the monoculture was treated principally due to their smaller size caused by intra-crop competition for available soil moisture (O'Donnell & Coaker, 1975).

The dense plant cover provided by the intercrop probably caused the reduction in aphid attack, the immigrant aphids preferring plants that stand out against a bare soil background. Plant cover, might, therefore, diminish the optomotor landing response of the migrating aphid with intercropping providing the appropriate level of camouflage. Fewer D. radicum eggs and damage were found on intercropped brassicas due to both a lower 'resource concentration' that increased the emigration rate of the adult flies from the plots caused by the diversionary effects of the non-host plants, and the enhanced predatory activities of carabids and staphylinids from the increased ground cover. Evidence from treatments in which beetles were excluded suggested that the diversionary effects were more important (Tukahirwa & Coaker, 1982). Knowledge of the mechanisms effecting pest suppression provides guidance for 'fine tuning' of the intercropping system. For brassicas, a good ground cover provided by the intercrop is important for aphid control, and comparable height and growth rates of both crops are necessary to divert female D. radicum away from ovipositing around the host plant. The correct choice of intercrop and its sowing time is, therefore, vital to the success of the system. The distance between the rows of the brassica and intercrop is also critical, since when more than 50 cm apart the levels of pest suppression are reduced, conversely the closer the rows are together the greater is the effect of intercrop competition.

1.2.2 Carrots

The gardener's practice of intercropping carrots with onions to suppress carrot fly, Psila rosae, damage by the supposed repellant effect of the onion odour was tested experimentally (Uvah & Coaker, 1984).

P. rosae has two main generations a year and the effect of intercropping carrots with onions on each generation was investigated. The treatments included a range of carrot/onion plant ratios obtained by intercropping single rows of carrots with 1, 2 and 4 rows of onions. Both crops were either sown at the same time in March or the carrots in March and the onions in June so that the effect of young versus senescing (bulbing) crops could be compared. Another pattern of mixed cropping was examined by growing onions at different densities around plots of carrots. This system tested the effects of onion odour on host plant finding by female P. rosae whereas the inter- row cropping could also have involved diversionary effects by the non-host crop on the egg laying activity of the flies (cf. D. radicum).

Whereas intercropping caused similar increases in carabid and staphylinid predator populations during both P. rosae generations, there were larger reductions in P. rosae immatures during the first than second suggesting that predation alone was unlikely to have been the cause of the difference unless predator effectiveness differed between the two generations (Burn, 1982). Losses of marked P. rosae eggs did not show, however, comparable trends with beetle numbers, suggesting that other factors caused the reduction of P. rosae immatures and damage in the intercropped treatments.

Greenhowe (1930) suggested that female P. rosae located the host plant by its "sense of smell", Brunel (1977) by its colour and Stadler (1977) by its leaf shape. Beruter and Stadler (1971) and Guerin and Stadler (1982) have more recently shown that P. rosae responds both neurophysiologically and behaviourally to host plant specific chemicals. In the intercropping experiments female P. rosae was more responsive to carrots than males, and more were trapped entering plots upwind suggesting odour modulated anemotaxis. The presentation of a range of carrot:onion density ratios, therefore, probably provided different levels of chemical apparency (Feeny, 1976) to the flies the lower carrot/onion ratios being less apparent. Fig. 1 shows the relationship between the responses of the flies expressed as the % reduction in infestation to a range of carrot/onion ratios compared with that on carrots in monoculture. The lower ratios of carrots:onions reduced egg numbers and subsequent larval damage suggesting that as the concentration of onion volatiles increased host plant finding by P. rosae was disturbed. The level of reduction of P. rosae infestation by intercropping was also different between generations and the pattern of mixed cropping used. For example, the onions were more effective oviposition inhibitors during the first than second generation, suggesting a greater disturbance from young than older onions. The young, growing onion plants probably emitted more inhibitory volatiles than plants starting to bulb. Onions arranged peripherally around carrot plots gave similar results to those intercropped with carrots suggesting that the onion volatiles were more important than the diversionary effect of the non-host plants in suppressing P. rosae attack.

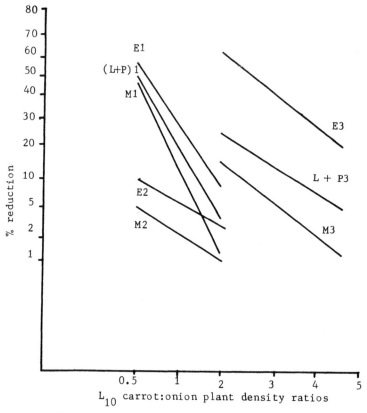

Fig. 1. Percentage reduction of P. rosae eggs (E), larvae plus pupae (L+P) and larval units (M) from monoculture plots with different carrot: onion density ratios. I - 1st generation (young onions), 2 - 2nd generation (senescing onions), 3 - 2nd generation (young onions).

The conflicting results obtained earlier by Petherbridge et al. (1942) and Hills (1972), from intercropping carrots with onions may, therefore, be explained by the timing of the carrot damage assessment, the later the assessment the less effective the intercropping became in reducing P. rosae damage, due to the ageing of the onions.

Intercropping carrots with onions as a cultural control method for P. rosae is unlikely to be effective on its own to provide the high level of damage control required for this crop (Wheatley, 1971). The system may, however, be useful on early carrot crops or to gardeners and growers concerned with limiting pesticide inputs.

1.3 Conclusions

Intracrop diversity could provide for some crops benefits over monocultures. Its integration with lower pesticide inputs, for example, necessary to produce high quality crops should reduce side-effects on beneficial species and delay the onset of resistance.

Against this cultural control system is the amount of agronomic effort required to produce finely tuned systems to maximise pest suppression and its conflict with current technology developed specifically for monocultural cropping. Intra-crop diversity could nevertheless provide cropping systems whose long-term benefits might outweigh some of the short-term economic costs of crop production.

REFERENCES

1. BERÜTER, J. and STADLER, E. (1971). An oviposition stimulant for the carrot fly from carrot leaves. Z.Naturf. 26, 339-340.
2. BURN, A.J. (1982). The role of predator searching efficiency in carrot fly egg loss. Ann.Appl.Biol. 101, 154-159.
3. BRUNEL, E. (1977). Etude de l'attraction periodique de femelles de Psila rosae par la plante-hote et influence de la vegetation environnante. Comportment des insectes et milieu Trophique. Colloques Internationaux du CNRS No. 265, 373-389.
4. COAKER, T.H. (1980). Insect pest management in brassicas by intercropping. IOBC/WPRS bulletin 1980/III/I: 117-125.
5. DEMPSTER, J.P. and COAKER, T.H. (1974). Diversification of crop ecosystems as a means of controlling pests. In: D. Price-Jones and M.E Solomon (eds.), Biology in Pest and Disease Control. Blackwell, Oxford, 106-114.
6. FEENY, P. (1976). Plant apparency and chemical defence. Biochemical interaction between plants and insects. Rec.Adv.Phytochem. 10, 1-40.
7. GREENHOWE, G.E. (1930). Carrot growing in gardens and allotments with special reference to control of carrot fly. Scott. J.Agric. 178-184.
8. GUERIN, P. and STADLER, E. (1982). Host odour perception in three phytophagous diptera - a comparative study. In: J.H. Visser and A.K. Minks (eds.) Proceedings of the 5th International Symposium on Insect and Plant Relationships, Wageningen, 1982. PUDOC Wageningen: 95-105.
9. HILLS, L.D. (1972). Pest control without poisons. Henry Doubleday. Res.Assn. Bocking, England.
10. LEVINS, R. and WILSON, M. (1980). Ecological thoery and pest management. Ann.Rev.Entomol. 11, 287-309.
11. MEAD, R. and WILLEY, R.W. (1980). The concept of a land equivalent ratio and advantages in yields from intercropping. Expl.Agric. 16, 217-228.

12. NORTON, G.A. and CONWAY, G.R. (1977). The economic and social context of pest, disease and weed problems. In: J.N. Cherrett and G.R. Sagar (eds.). The Origins and Pests, Parasites, Disease and Weed Problems. Brit.Ecol. Soc. Symposium, 205-226.

13. O'DONNELL, M.S. and COAKER, T.H. (1975) Potential of intracrop diversity in the control of brassica pests. Proc.8th Brit.Insectic. Fungic.Conf. 1975, 101-105.

14. PETHERBRIDGE, F.R.D., WRIGHT, D.W. and DAVIES, P.G. (1942) Investigations on the biology and control of carrot fly (Psila rosae). Ann.Appl.Biol. 29, 380-392.

15. PERRIN, R.M. (1977). Pest management in multiple cropping systems. Agro-ecosystems 3, 93-118.

16. PRICE, P.W. and WALDBAUER, G.P. (1975). Ecological aspects of insect pest management. In: R.L. Metcalf and W.H. Luckman (eds.) Introduction to Insect Pest Management. Wiley, New York, 36-73.

17. RISCH, S.J., ANDREW, D. and ALTIERI, M.A. (1983). Agroecosystem diversity and pest control: data, tentative conclusions and new research directions. Environ.Entomol. 12, 625-629.

18. ROOT, R.B. (1973). Organisation of a plant-arthropod association in simple and diverse habitats: the fauna of collards (Brassica oleracera). Ecol. Monogr. 43, 74-125.

19. STADLER, E. (1977). Host selection and chemoreception in the carrot rust fly (Psila rosae, F.): extraction and isolation of the oviposition stimulants and their perception by the female. Comportments des insectes et milieu Trophique. Colloques Internationeaux du CNRS. No. 265, 357-372.

20. TUKAHIRWA, E.M. and COAKER, T.H. (1982). Effects of mixed cropping on some insects pests of brassicas; reduced Brevicoryne brassicae infestations and influences on epigeal predators and the disturbance of ovipositional behaviour in Delia brassica. Entomol.exp. appl. 32, 129-140.

21. THEUNISSEN, J. and OUDEN, H. Den (1980). Effects of intercropping with Spergula arvensis on pests of Brussels sprouts. Entomol.exp. appl. 27, 260-268.

22. EMDEN, H.F. VAN and WILLIAMS, G.F. (1974). Insect stability and diversity in agroecosystems. Ann.Rev.Entomol. 19, 455-475.

23. UVAH, I.I.I. and COAKER, T.H. (1984) Effect of mixed cropping on some insect pests of carrots and onions. Entomol.exp.appl. 36, 159-167.

24. WILLEY, R.W. (1979). Intercropping - its importance and research needs Parts 1 and 2. Field Crop Abst. 32, 1-10 and 73-85.

25. WHEATLEY, G.A. (1971). The role of pest control in modern vegetable production. World Rev.Pest Control 10, 81-93.

Present status of supervised pest control methods in field vegetables

J.Freuler
Federal Agricultural Research Station of Changins, Nyon, Switzerland

Summary

The present status of observation and field sampling methods suitable
for practising supervised insect pest control in field vegetables is
analysed. The methods are classified according to procedure criteria
and for the 20 insects dealt with the availability of the kind of
forecast (negative prevision or control thresholds) is indicated.
The criteria of the acceptability of a method for practical application
are evoked.
For the majority of the insect pests the easiest methods are offered
to be applied by the growers themselves.

Résumé

La situation actuelle des méthodes de contrôle utilisables pour prati-
quer la lutte dirigée contre les ravageurs des cultures maraîchères de
pleine terre est analysée.
Les méthodes sont classées selon des critères de procédure et pour les
20 ravageurs traités la disponibilité du type de prévision (prévision
négative ou seuil) est indiquée.
Les critères d'acceptabilité d'une méthode par la pratique sont évo-
qués. On constate que pour la majorité des ravageurs les méthodes les
plus simples sont offertes et peuvent être exécutés par les maraîchers
eux-mêmes.

Many efforts have been made during these last years to promote integrated pest management in field vegetables. This concept is still far from
being reality for these annual crops. To reach the ideal of plant
protection a slow approach by steps is generally chosen, whereby the first
step is considered to be supervised pest control.
This means in a practical way that one needs some kind of field
sampling in order to be able to evaluate the pest situation in a field. A
decision has then to be taken whether or not and, if possible, how control
measures have to take place.
This paper presents a collection of observation and field sampling
methods to be applied for insect pests attacking field vegetables.
According to the pest target these observations carried out following
the preconized methods can
- produce final results or give rise to ensuing observations
- start within the crop or outside the crop
- terminate on the field or in the laboratory
- need or not special devices (e.g. traps, apparatus for extraction)

Table 1a. Observation methods available for 20 insect pests in field vegetables.

Producing final results							
Starting within the crop							
Terminating on the field				In the laboratory			Outside the crop
Without material			With material		Without material	With material	
Related to insect pest	Damage	Condition host plant	Related to insect pest	Damage	Damage	Related to insect pest	Damage
Agrotis ipsilon HFN.[1]	Agrotis ipsilon HFN.[1]	Sitona lineatus L.[1]	Psila rosae Fab.[1+2]			Agriotes spp.[2]	
Agriotes spp.[1]	Napomyza carotae SPENCER[1]		Delia radicum L.[1]			Tipula spp.[2]	
Trioza apicalis FORST.[1]	Liriomyza nietzkei SPENCER[1]		Platyparea poecioptera SCHR.[1]				
Thrips tabaci LIND.[1]	Ceuthorrhynchus suturalis FABR.[1]						
Mamestra brassicae L.[1]	Delia platura MEIG.[1]						
Caterpillars on cabbage[1]	Sitona lineatus L.[1]						
Brevicoryne brassicae L.[1]							
Athalia rosae L.[1]							
Aleyrodes proletella L.[1]							
Pemphigus bursarius L.[1]							
Delia platura MEIG.[1]							
Aphis fabae SCOP.[1]							

[1] Observation which can be made by the grower
[2] Observation made by the extension service

Table 1b.

Giving rise to ensuing observations									
Starting within the crop					Outside the crop				
Terminating on the field				In the laboratory	Terminating on the field			In the laboratory	
Without material		With material			Without material		With material	Without material	With material
Related to insect pest	Damage	Related to insect pest	Damage		Related to insect pest	Damage	material	material	Related to insect pest	Damage
Pemphigus bursarius L.[1]		Mamestra brassicae L.[1]			Pemphigus bursarius L.[2]				Agrotis ipsilon HFN.[2]

[1] Observation which can be made by the grower
[2] Observation made by the extension service.

291

Table 2. Availability of forecast for insect pests in field vegetables.

Insect pest	Availability of forecast		Remark
	Negative prevision	Tolerance level	
Miscellaneous crops			
Agrotis ipsilon HFN.	x	(x)	Provisional tolerance level. Risk estimation on a regional scale.
Agriotes spp.	x	(x)	Provisional tolerance level depending on a great number of factors.
Tipula spp.	x	(x)	Provisional tolerance level.
Carrot			
Psila rosae FAB.		x	Variable tolerance level according to growing area and sowing date.
Napomyza carotae SPENCER	x	x	Tolerance level worked out for Germany.
Trioza apicalis FORST.	x		
Oignon and leek			
Liriomyza nietzkei SPENCER		x	
Ceuthorrhynchus suturalis FABR.	x		
Thrips tabaci LIND.	x	x	Tolerance level for oignon. Pheromone trap not reliable.
Acrolepiopsis assectella Z.			
Cruciferous crops			
Delia radicum L.	x	(x)	Provisional tolerance level for cabbage.
Mamestra brassicae L.	x	x	Tolerance level worked out for Germany.
Caterpillars on cabbage (Mamestra brassicae L.*, Pieris* spp.*, Autographa gamma* L.*, Plutella xylostella* L. & *Evergestis forficalis* L.*)*		x	Tolerance level worked out for The Netherlands and Germany.
Brevicoryne brassicae L.		x	Tolerance level worked out for The Netherlands and Germany.
Athalia rosae L.	x		
Aleyrodes proletella L.		x	Tolerance level worked out for Germany.

Table 2 (continued).

| Insect pest | Availability of forecast | | Remark |
	Negative prevision	Tolerance level	
Lettuce and Witloof			
Pemphigus bursarius L.	x		
Asparagus			
Platyparea poeciloptera SCHR.	x		
Bean and pea			
Delia platura MEIG.	x	(x)	Provisional tolerance level.
Sitona lineatus L.	x	x	Tolerance level worked out for Germany.
Aphis fabae SCOP.	x	x	Tolerance level worked out for Germany.

- be related to the insect pest, to the damage or to the condition of the host plant.

These different observation types are shown in table 1. The 20 different insect pests are put in the place which correspond to the method to be applied for its case. Some insects appear at several places because various observation methods are suitable, so that one arrives at 28 cases.

For 19 of them the simpliest method is proposed, that means the observation produces final results, starts within the crop, terminates in the field and does not need special devices. In 12 cases the insect pest is concerned, in 6 cases the damage and in one case the condition of the host plant.

In general an observation method can reach two signification levels: qualitative or quantitative.

In the first case, the observation elements indicate whether or not an insect pest is present in the observation area. This information serves then for what is commonly called negative prognosis.

The second case is used if working with control thresholds. For some insect pests and crops variable tolerance levels are specified in order to take into account the period of the crop during the growing season and the growth stage of the crop (1).

Table 2 summarizes the present situation concerning the prognosis of insect pests in field vegetables. One notices that for many cases no threshold is yet available and that for Acrolepiopsis assectella the situation is particularly unsatisfactory.

Some difficulties check the development of observation methods in field vegetables. Amongst these is to be mentioned:
- sampling methods in high plant density crop
- soil born insect pests
- dispersing or migrating insect pests.

The details of the observation and field sampling methods as summarized in tables 1 and 2 are described earlier (2).

Observation methods are often worked out by entomologists in order to study population dynamics of an insect pest. At this stage a good precision and reliability of the method is essential to get the maximum of basic information. Once the method has been proved effectual one has to think of how to simplify the procedure in order to render it acceptable to the grower. Now, if the datas on table 1 are reconsidered it can be noted that the simplification is in an advanced stage.

It is therefore useful to define clearly what is an acceptable observation method for practice.

Two aspects have to be considered. First, there is the length of time which is needed for an observation. This is a calculable economic factor when considering the value of the crop and the possible savings of production costs (direct costs =(pesticide) treatment or indirect costs = negative effects on parasites and predators and on the environment).

Second, there is the complexity of the methods. This is a training problem which is to be overcome but which in most cases goes beyond the activities of the people involved in developing these procedures.

As an indication, table 1 shows for which insect pest the observation can be accomplished by the grower or has to be carried out by extension services or eventually in common with the grower.

The educational standard of the profession in crop production is steadily increasing and it certainly can be raised in plant protection, too. Anyway, the complexity of a method must not be a reason to give it up beforehand because it is finally the phytosanitary situation which takes precedence over everything: if the traditional pest control methods continue to be cheap and guarantee a certain security, any method as simple as it may be will have little chance to be adopted.

References

1. THEUNISSEN, J.(1984).Supervised pest control in cabbage crops: theory and practice. Mitt. Biol. Bundesanst. Land- Forstwirtsch. Berlin-Dahlem H. 218, 76-84.
2. FREULER, J. & FISCHER, S.(1985).Méthodes de contrôle et utilisation des seuils de tolérance pour les ravageurs des cultures maraîchères de pleine terre. Revue suisse Vitic. Arboric. Hortic. 17 (4): 227-246.

Conclusions and recommendations

Introduction

The meeting of experts on "Integrated plant protection in field vegetables" was organised jointly by the Commission of the European Communities and the International Organisation for Biological and integrated Control.

The most important aspects concerned the use of less pesticides in vegetable growing, and the urgency to investigate alternative methods to control the main pests and diseases.

This aim became clear when the investigations of researchers were compared.

The experts presented their results on crops of brassica, carrots, tomatoes, potatoes, beans and artichokes.

All the presentations were followed by open and fruitful discussions, in view of reaching effective information which can be used in agricultural practice.

Conclusions

The aspects which emerged most clearly, and which require consideration in the future, may be summarised as follows:

. for brassica crops (Belgium, Denmark, France, Germany, Great Britain, Holland, Ireland)

- the main pests include the Diptera and the Lepidoptera, and the diseases that are due to Plasmodiophora;
- the need for more knowledge on parasites and predators emerged strongly, in view of their more rational use in practice;
- the elements to be considered for better protection are trapping, monitoring, and the possibilities offered by a presence-absence warning for the development of the threshold concept.

. for carrot crops (Denmark, France, Great Britain, Holland, Switzerland)

- greater knowledge of the biology of Psila rosae, and in particular of its behaviour, dynamics and population fluctuations, is necessary;

- trapping methods for the carrot fly should be seriously taken into account;
- other pests to be considered are _Agrotis segetum_ and _Heterodera carotae_;
- among the significant points considered, a definite interest was shown in the possibilities offered by the use of volatile substances, either attracting or repelling, applied directly, as well as the importance of the utilization of intercalated crops.

. for tomato crops (France, Greece, Italy, Spain)

- the problems caused by pests such as Lepidoptera, Aphids and Mites emerged;
- the validity of biological control was stressed;
- the importance of resistance to diseases was mentioned;
- the bacterial disease status was revised.

. for other horticultural crops (Greece, Italy, Portugal)

- the importance of Aphids for the transmission of viral diseases to potatoes;
- the susceptibility to bacterial diseases of different varieties of beans;
- the validity of the integrated control method against Lepidoptera which are dangerous in artichoke crops.

Recommendations

The experts greatly appreciated the opportunity to meet, and learn about the different problems which are posed in the different countries concerning the application of integrated control to vegetable crops in the open field.

The experts hope that further such meetings may be organised, given the undeniable interest in the application of alternative methods of protection which lead to the production of healthy vegetables, which are hardly or not at all treated with toxic chemical products.

The participants think that exchanges of researchers, and the establishment of joint programme, could speed up a wide-ranging application of the results. To achieve this goal they propose that certain research themes be pursued in close collaboration.

The participants also hope that young researchers be given the possibility of specialising in the various aspects of integrated pest control by spending research time in specialised laboratories.

The experts believe that harmonisation of phytosanitary regulation is greatly to be desired, and feel that the acceptance of protocols established for the examination of biopesticides, and the promulgation of EC directives concerning their registration, will encourage industry to commercialise these useful products.

List of participants

Belgium

       VAN KEYMEULEN Martine
       IWONL-Centrum voor Geintegreerde Bestrijding
       Fakulteit van de Landbouwwetenschappen
       Coupure Links, 653
       9000 - Gent

Denmark

       BROMAND Bent
       Research Centre for Plant Protection
       Institute of Pesticides
       Lottenborgvej, 2
       2800 - Lyngby

       ESBJERG Peter
       Research Centre for Plant Protection
       Zoological Department
       Lottenborgvej, 2
       2800 - Lyngby

       PHILIPSEN Holger
       Royal Veterinary and Agricultural University
       Department of Zoology
       Bulowsvej, 13
       1870 - Frederiksberg

France

       BOSSIS Michel
       I.N.R.A. - Laboratoire de Zoologie
       Domaine de la Motte-au-Vicomte - B.P. 29
       35650 - Le Rheu

       BRUNEL Etienne
       I.N.R.A.
       Domaine de la Motte-au-Vicomte - B.P. 29
       35650 - Le Rheu

BUES Robert
I.N.R.A. – Station de Zoologie
Domaine St. Paul
84310 – Montfavet

POITOUT Serge
C.R.A. d'Avignon
Station de Zoologie et d'Apidologie
Domaine St. Paul
84310 – Montfavet

RAHN Robert
I.N.R.A. – Laboratoire de Zoologie
Domaine de la Motte-au-Vicomte
35650 – Le Rheu

ROUXEL Francis
I.N.R.A. – Station de Pathologie Végétale
Domaine de la Motte-au-Vicomte
35650 – Le Rheu

F.R. Germany

HOMMES Martin
Federal Biological Research Centre
for Agriculture and Forestry
Institute for Plant Protection
in Horticultural Crops
Messeweg, 11/12
3300 – Braunschweig

Greece

PANAGOPOULOS Christos
Athens College of Agricultural Sciences
Laboratory of Plant Pathology
75, Iera Odos
11855 – Athens

Ireland

DUNNE Richard
An Foras Taluntais
Kinsealy Research Centre
Malahide Road
Dublin 17

Italy

CIRULLI Matteo
Dipartimento di Patologia Vegetale – Università
Via Amendola, 165-A
70100 – Bari

TREMBLAY Ermenegildo
Istituto di Entomologia Agraria
Università di Napoli
Via Università, 100
80055 - Portici

Netherlands

DEN OUDEN Hugo
Instituut Plantenziektenkundig Onderzoek
Binnenhaven, 12 - P.O. Box 42
6700 - Wageningen

DE PONTI Orlando
Institute for Horticultural Plant Breeding
P.O. Box 16
6700 - Wageningen

THEUNISSEN Jan
Research Institute for Plant Protection
Binnenhaven, 12
6700 - Wageningen

Portugal

CRUZ-DE BOELPAEPE Maria Odilia
Centro Nacional de Proteccao
de Producao Agricola
Tapada da Ajuda
1300 - Lisboa

Spain

ALBAJES Ramon
Institut d'Investigacio
i Desenvolupament agrari de Lleida
Universitat Politecnica de Catalunya
Proteccio de Conreus
Rovira roure, 177
25006 Lleida

Suisse

FISCHER Serge
Station Fédérale de Recherches Agronomiques
de Changins
1260 - Nyon

FREULER Jost
Station Fédérale de Recherches Agronomiques
de Changins
1260 - Nyon

KELLER Karin
Eidgenossische Forschungsanstalt fur Obst-
Wein- und Gartenbau Schloss
8820 - Wadenswil

TERRETTAZ Catherine
Station Cantonale d'Agriculture
1950 - Chateauneuf

United Kingdom

BIRCH Nicholas
Scottish Crop Research Institute
Village Invergowrie
Dundee DD22 5DA

COAKER T.H.
Department of Applied Biology
University of Cambridge
Pembroke Street
Cambridge CB3 9LU

COLLIER Rosemary
Zoology Department
University of Birmingham
Birmingham B1S 2TT

ELLIS Peter Robin
National Vegetable Research Station
Wellesbourne
Warwick CV35 9EF

FINCH Stan
National Vegetable Research Station
Wellesbourne
Warwick CV35 9EF

HUMPHREYS Ian Clive
Department of Agricultural Zoology
Queen's University of Belfast
Newforge Lane
Belfast BT9 5PX
Northern Ireland

MOWAT D.J.
Department of Agriculture for Northern Ireland
and Queen's University of Belfast
Newforge Lane
Belfast BT9 5PX
Northern Ireland

SKINNER Gareth
National Vegetable Research Station
Wellesbourne
Warwick CV35 9SB

International Organizations

C.E.C.

      CAVALLORO Raffaele
      Commission of the European Communities
      Joint Research Centre
      I - 21020 Ispra

I.O.B.C.

      PELERENTS Christian
      Faculty of Agricultural Sciences
      Coupure Links 653
      B - 9000 Gent

Index of authors

ALBAJES, R. 197
ALOMAR, O. 197

BIRCH, N. 123
BIRIS, D.A. 275
BORDAS, E. 197
BOSSIS, M. 187
BRIARD, M. 145
BROMAND, B. 15, 33, 75
BRUNEL, E. 33
BUES, M. 33
BUES, R. 209

CARNERO, A. 197
CASTAÑE, C. 197
CAVALLORO, R. 3
CICCARESE, F. 223
CIRULLI, M. 223
COAKER, T.H. 281
COFFEY, J. 95
COLLIER, R.H. 21, 27, 33, 37, 49
CRUZ DE BOELPAEPE, M.O. 257

DEN OUDEN, H. 11, 83, 117, 161
DUNNE, R. 33, 95

ELLIS, P.R. 99
ESBJERG, P. 177

FINCH, S. 21, 27, 33, 37, 45, 49, 61,
  155
FOSTER, G. 33
FREULER, J. 33, 135, 289

GABARRA, R. 197
GFELLER, F. 135, 167

HARDMAN, J.A. 99
HOMMES, M. 33, 55, 135

KELLER, K. 167

LEJEUNE, B. 145

MOWAT, D.J. 33, 87, 135

PANAGOPOULOS, C.G. 221, 275
PELERENTS, C. 7, 33, 67, 129
PHILIPSEN, H. 167, 169
POITOUT, H.S. 209

RAHN, R. 137
RODRIGUES, M.I. 257
ROTUNDO, G. 277
ROUXEL, F. 145

SKINNER, G. 33, 45, 61, 155
STÄDLER, E. 33, 167

THEUNISSEN, J. 33, 107, 117, 135, 161
TOUBON, J.F. 209
TREMBLAY, E. 277

VAN DE VEIRE, M. 129
VAN DER STEENE, F. 135
VAN KEYMEULEN, M. 33, 67